T0344680

Advances in Bioceramics and Biotechnologies II

Advances in Bioceramics and Biotechnologies II

Ceramic Transactions, Volume 247

A Collection of Papers Presented at the
10th Pacific Rim Conference on
Ceramic and Glass Technology
June 2–6, 2013
Coronado, California

Edited by
Joanna M. McKittrick
Roger Narayan

Volume Editor
Hua-Tay Lin

The
American
Ceramic
Society

WILEY

Published by John Wiley & Sons, Inc., Hoboken, New Jersey.
Published simultaneously in Canada.

For general information on our other products and services or for technical support, please contact our
Customer Care Department within the United States at (800) 762-2974, outside the United States at
(317) 572-3993 or fax (317) 572-4002.

Wiley also publishes its books in a variety of electronic formats. Some content that appears in print may
not be available in electronic formats. For more information about Wiley products, visit our web site at
www.wiley.com.

Library of Congress Cataloging-in-Publication Data is available.

ISBN: 978-1-118-77139-6
ISSN: 1042-1122

Printed in the United States of America.

10 9 8 7 6 5 4 3 2 1

Contents

**NANOSTRUCTURED BIOCERAMICS AND CERAMICS FOR
BIOMEDICAL APPLICATIONS**

Preface

Continued discoveries and innovations have resulted in novel nanostructured ceramics for medical and dental applications. For example, novel processing, characterization, and modeling approaches for developing nanostructured ceramics are becoming available. In addition, nanostructured ceramics may enable improved interactions with proteins and other biological molecules. The Nanostructured Bioceramics and Ceramics resulted in lively discussions and interactions among the many groups involved in the development and use of bioceramics, including ceramic researchers, medical device manufacturers, and student researchers.

This Ceramic Transactions volume represents selected papers based on presentations in two symposia during the 10th Pacific Rim Conference on Ceramic and Glass Technology, June 2–6, 2013 in Coronado, California. The symposia include:

• Advances in Biomineralized Ceramics, Bioceramics, and Bioinspired Designs
• Nanostructured Bioceramics and Ceramics

The editors wish to extend their gratitude and appreciation to all the co-organizers for their help and support, to all the authors for their cooperation and contributions, to all the participants and session chairs for their time and efforts, and to all the reviewers for their valuable comments and suggestions. Thanks are due to the staff of the meetings and publication departments of The American Ceramic Society for their invaluable assistance. We also acknowledge the skillful organization and leadership of Dr. Hua-Tay Lin, PACRIM 10 Program Chair.

We hope that this proceedings will serve as a useful resource for the researchers and technologists in the fields of bioceramics, nanobiomaterials, and biomineralization.

JOANNA M. MCKITTRICK, University of California, San Diego
ROGER NARAYAN, University of North Carolina and North Carolina State
   University

Advances in Biomineralized Ceramics, Bioceramics, and Bioinspired Design

VAPOR DEPOSITION POLYMERIZATION AS AN ALTERNATIVE METHOD TO ENHANCE THE MECHANICAL PROPERTIES OF BIO-INSPIRED SCAFFOLDS

Pei Chun Chou[1], Michael M. Porter[2], Joanna McKittrick[2,3], Po-Yu Chen[1*]

1 Department of Material Science and Engineering, National Tsing Hua University, Hsinchu, ROC
2 Materials Science and Engineering Program, University of California, San Diego, La Jolla, CA 92093, USA
3 Department of Mechanical and Aerospace Engineering, University of California, San Diego, La Jolla, CA 92093, USA
* Corresponding author, email: poyuchen@mx.nthu.edu.tw

ABSTRACT

Cancellous bone is a composite of a biopolymer phase (type-I collagen) and a mineral phase (carbonated hydroxyapatite) assembled into a hierarchically structured, highly-porous network, which possesses proper mechanical strength. In this study, two types of ceramic-based scaffolds were synthesized to mimic the structure and mechanical performance of cancellous bone. Natural scaffold was obtained by complete deproteinization of bovine cancellous bone while synthetic scaffold of alumina was prepared by freeze casting. Both types of ceramic scaffolds were then infiltrated with a polymer phase (PMMA) by vapor deposition polymerization. Microstructural features of bio-inspired scaffolds were characterized by stereoscopy, scanning electron microscopy (SEM) and micro-computed tomography (μ-CT). Compressive tests showed that vapor deposition polymerization can successfully enhance the mechanical performance of scaffolds. Microstructural features and mechanical properties of scaffolds can be tunable by controlling the amount of monomer, cooling rate as well as grafting and annealing treatments. The compressive strength of scaffolds increased with increasing cooling rate and the amount of monomer applied. Toughening mechanisms at the ceramic/polymer interface, such as crack deflection, uncracked ligament bridging and microcrack formation were observed and discussed.

INTRODUCTION

Natural hard tissues, such as bone and teeth, are composites of mineral and organic constituents organized into complex, hierarchical structures which possess exceptional mechanical properties [1-4]. Bio-inspired synthesis of ceramic-based scaffolds has attracted more and more attention due to the potential applications in the field of light-weight composites and biomedical materials, such as bone filler or substitutes [5-9]. One of the most promising novel techniques to synthesis porous ceramic scaffolds is by freeze casting, a physical procedure where an aqueous slurry of solid particles (e.g. ceramic particle) and liquid carrier is directionally frozen then followed by removing the frozen liquid phase, sintering and/or infiltrating a second phase (e.g. polymer). Many research groups have been working in this area and had developed unique methods to fabricate scaffolds or composites with various structure, properties and functionalities for a wide range of applications [10-15]. However, an inherent drawback that limits the application of freeze casted scaffolds is their lack of strength and stiffness. The infiltration of an organic phase into an inorganic scaffold is a proven method to enhance both the strength and toughness of such scaffold [15]. Common infiltration methods include polymer melt immersion [16], polymer-solvent evaporation [17], in-situ polymerization [10,18], centrifugation [19], and vapor deposition polymerization [20]. Polymer melt immersion and polymer solvent evaporation are similar to the dip coating [16,17]. Porous scaffolds are immersed in a polymer melt or solvent to allow the infiltration of the polymer phase. Thin film of polymer is then coated onto the scaffolds when the ceramic scaffolds are pulled out from the polymer medium. In-situ polymerization used monomer solution for infiltration [10,18]. Once the monomer solution completely infiltrates into the scaffolds, initiator and heat are applied to trigger polymerization. Centrifugation is a slightly different technique which use solid polymer particle as infiltration medium [19]. With the assist of centrifugation force, polymer particles which suspended in water are forced to infiltrate into the porous scaffolds. However, the above-mentioned polymer infiltration techniques tend to completely fill the porosity in the scaffolds and are not suitable for bone grafting and repairing. Vapor deposition polymerization has the potential to control the porosity of the composite by changing the process time and the amount of monomer being used. In this study, vapor deposition polymerization is utilized to enhance the mechanical performance of ceramic scaffolds.

EXPERIMENTAL METHODS
Scaffold Preparation
Ceramic scaffolds are prepared by two different methods. Natural mineral scaffolds were obtained by complete deproteinization of cancellous bovine femur following the procedure described by Chen et al. [21-23]. Mature bovine femur bone was purchased from a local market. Soft tissues attached to the femur were removed by hands and cutting knife. The cancellous bone was obtained from the proximal femur head and sectioned into rectangular blocks (~ 5 x 5 x 7.5 mm^3) with a rotating diamond blade. Bone marrow and blood that remain within the cancellous bone were carefully cleaned by deionized (DI) water and compressed gas. Ten cancellous bone samples were air-dried and kept untreated before mechanical testing, while the other ten samples were immersed in a bleach solution (~3 wt.% NaOCl) for deproteinization. The deproteinization continued for 8 days and the solution was held at 37°C to accelerate the process. The bleach solution was replaced daily to ensure that fresh NaOCl was available all the time. Deproteinized samples was then stored in a desiccator after being rinsed with DI water and air dried.

Following the similar process described by Porter et al. [13], freeze casting was performed by a custom-made freeze casting unit which consists of a copper metal finger, a PVC mold, a ring heater that attached to a PID controller, and a Styrofoam tank to temporary store liquid nitrogen. Alumina powders (20 vol.%, Sigma Aldrich, St. Louis, MO) were first mixed with organic binder, PVA and PEG (~1 wt.%, Alfa Aesar, Ward Hill, MA) and anionic dispersant (~1 wt.% , DARVAN® 811 R. T. Vanderbilt Co., Norwalk, CT), then ball milled for 24 hours. About 20 ml of slurry was poured into the PVC mold and frozen at the cooling rate of 2°C/min or 10°C/min. After freezing, the samples were removed from the mold with hydraulic pressure and lyophilized in a freeze dryer at – 40°C and 128Pa for 72 hours. The freeze-dried green body was then transferred into an open-air furnace for sintering. The firing schedule was set to initially heat up to 600°C to remove the organic binder then heat up to 1400°C with the heating rate of 2°C/min. Temperature is held at 1400°C for 8 hours to ensure complete sintering. The fired body was carefully sectioned to the dimension of 5 x 5 x 7.5 mm^3 and stored in desiccator.

Polymer Infiltration
Both types of scaffolds were infiltrated with polymethylmethacrylate (PMMA) by the vapor deposition polymerization (VDP) process. Special designed glass-made tube that contains two chambers was used for the vapor deposition process. First, 0.5 or 1 ml of methylmethacrylate (MMA, Sigma Aldrich) solution was carefully injected into the second chamber. Porous scaffold is then transferred into the first chamber along with 0.05g of the initiator, AIBN (Sigma Aldrich). Next, the tube was evacuated to 10^{-3} Pa with a mechanical pump, and then carefully sealed by a

torch. Finally, the sealed glass tube was transferred into an oven which was held at 70°C for 7 days.

Grafting and Annealing
To improve the interfacial adhesion between the inorganic scaffold and the organic polymer coating, grafting material, 3-(trimethoxysilyl)propyl methacrylate (γ-MPS, Alfa Aesar), was used. Grafting was carried out by soaking the porous scaffold into grafting agent (50 vol% acetone, 50 vol% γ-MPS) for at least 12 hours. Annealing was carried out after the vapor deposition polymerization is completed. Samples remained in the oven at an elevated temperature of 130°C for 24 hours.

Compressive Mechanical Testing
Samples were tested under compression at a crosshead speed of 0.3mm/min (equal to a strain rate of ~1×10^{-3}/sec) by a universal testing machine (Instron 3343 Single Column Testing System, Instron, Norwood, MA, USA) equipped with a 1kN load cell.

Structural Characterization
Stereoscopic images were taken from an Olympus SZX7 Zoom Stereomicroscope (Olympus Co., Tokyo, Japan) with a 2.0 megapixel CCD camera (Infinity 1, Lumenera Co., Ontario, Canada). Micro-structural characterization of the polymer infiltrated scaffold was carried out by scanning electron microscopy (SEM) and micro-computed tomography (μ-CT). Samples prepared for SEM observation were sectioned by diamond blade to produce cross-sectional surfaces. Samples were then polished and sputter-coated with a thin layer carbon. The fracture surfaces after compression tests were observed under SEM to analyze the fracture mechanism. SEM images were taken by using a field-emission scanning electron microscope (JEOL-7600, JEOL Ltd., Akishima, Tokyo, Japan). Micro-CT scans were accomplished by a synchrotron hard X-rays facility at the National Synchrotron Radiation Research Center (NSRRC) in Hsinchu, Taiwan. The images were captured by a CCD camera after converting the X-rays into visible lights by a scintillator. Xradia software (Xradia Inc., Pleasanton, CA, USA) was used to reconstruct the 3-dimensioal image from the raw data. The reconstructed images were then visualized by Amira software (Visualization Science Group, FEI co., Burlington, MA, USA).

RESULTS AND DISCUSSION
Structural Characterization of the Polymer Infiltrated Scaffold
Figure 1a shows the cross-sectional stereoscopic image of the completely deproteinized (DP) bone without polymerization. The DP cancellous bone maintains structural integrity with complex porous structure unaltered. PMMA infiltrated DP bone samples are shown in Figures 1b-d. Figures 1b and 1c show PMMA infiltrated DP bone samples without grafting and annealing, while Figure 1d demonstrate samples treated with grafting and annealing procedures. The preliminary results, as shown in Figure 1b and 1c, indicate that it is difficult to deposit a uniform, continuous layer of PMMA on the DP bone by VDP. Figure 1b shows that the polymer phase forms granular particles and clusters instead of uniform film and does not adhere well to the surface of DP bone. Figure 1c shows that water vapor and gas bubbles trapped in the polymer layer during the polymerization process. Both water vapor and gas present as the second phase during the polymerization process which may significantly weaken the mechanical properties of the scaffold. Figure 1d shows that both vapor inclusion and lack of interfacial adhesion can be improved by grafting and annealing. The DP bone is coated with a continuous layer of PMMA.

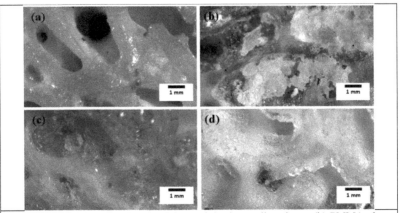

Figure 1. (a) Stereoscopic image of deproteinized cancellous bone. (b) PMMA phase forms granular particles and clusters in DP cancellous bone infiltrated by vapor deposition polymerization (VDP) without annealing and grafting. (c) Water vapor and gas bubbles trapped in the polymer layer during the VDP process. (d) Polymer phase forms a continuous film on the surface of DP bone after annealing and grafting treatment.

Figure 2a and 2b are SEM micrographs showing the microstructure of the freeze-casted scaffolds under two different cooling rates, 10°C/min and 2°C/min, respectively. The channel size and wall thickness of the scaffold can be tuned by adjusting the cooling rate of the slurry. As the freezing rate increases, the amount of supercooling ahead of the solidifying interface is increased, resulting in a finer microstructure. The scaffold synthesized at the cooling rate of 10°C/min has a finer and well-aligned microstructure with smaller channel spacing in the range of 10~20 μm, as shown in Figure 2a. When the cooling rate is reduced to 2°C/min, a relatively porous structure with large channel spacing (~100 μm) can be observed in Figure 2b.

The degree of polymer infiltration in the scaffold can be observed under SEM using the backscattered electron (BEI) mode. Figure 2c and 2d are BEI images showing the cross-sectional surfaces of the PMMA infiltrated alumina scaffolds. The darker region represents the PMMA phase while the brighter region represents the alumina scaffold. It is demonstrated that vapor deposition polymerization can achieve high degree of infiltration. However, some micron-sized pores and cavities can be observed in Figure 2d. This observation indicates that as the channel size increases, it becomes more difficult for the scaffold to be completely infiltrated by the vapor deposition process.

Figures 3a-d show micro-CT images of four types of PMMA infiltrated alumina scaffolds are synthesized in different conditions: (a) 10°C/min cooling rate, 1ml MMA; (b) 2°C/min cooling rate, 1ml MMA; (c) 10°C/min cooling rate, 0.5ml MMA; (d) 2°C/min cooling rate, 0.5ml MMA. The porosity is estimated based on the micro-CT images reconstructed by the Xradia software. Due to the different capability to absorb x-ray, the alumina scaffold and the PMMA coating reveal different grey-level on the images while the pores remains black. The scaffolds synthesized at 10°C/min cooling rate (Figures 3a and 3c) have finer microstructure compared with those synthesized at 2°C/min cooling rate (Figures 3b and 3d). The porosity increases with decreasing amount of MMA monomer. It can be observed in Figure 3a and 3c that porosity increases from 36% to 51% when the amount MMA solution being used decreases from 1ml to

0.5ml. In Figure 3b and Figure 3d, the porosity does not change significantly but large cavities (black region) can be observed in Figure 3d.

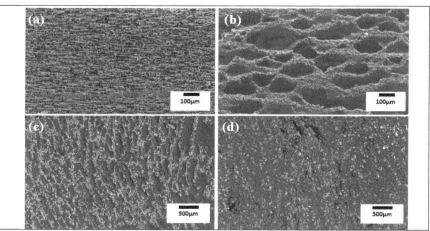

Figure 2. Cross-sectional backscattered electron (BSE) SEM images showing alumina scaffolds synthesized at (a) 10°C/min and (b) 2°C/min cooling rates, respectively. BSE images of PMMA infiltrated alumina scaffolds synthesized at (c) 10°C/min and (d) 2°C/min cooling rate, respectively.

Figure 3. Cross-sectional micro-CT sections and 3-dimensional models reconstructed by Xradia software (bottom left insets) showing microstructural features of PMMA infiltrated alumina scaffolds synthesized in different conditions: (a) Cooling rate 10°C/min, 1ml MMA, Porosity: 36% (b) Cooling rate 2°C/min, 1ml MMA, Porosity: 51% (c) Cooling rate 10°C/min, 0.5ml MMA, Porosity: 51% (d) Cooling rate 2°C/min, 0.5ml MMA, Porosity: 53%.

Mechanical Properties and Deformation Mechanisms

Compressive stress-strain curves are obtained from the tests and used to determine the Young's modulus and ultimate compressive strength of each sample. The results are summarized in Table I for natural scaffolds and Table II for alumina scaffolds.

Table I Mechanical Properties of PMMA Infiltrated Deproteinized (DP) Cancellous Bone Synthesized by Vapor Deposition Polymerization

	Young's modulus, E (MPa)	Ultimate compressive strength, σ (MPa)
#1 Grafting Only	1334±457	35±8
#2 Grafting + Annealing	1139±92	26±6
#3 Annealing Only	843±337	18±9
Deproteinized Bone (Scaffold)	316±38	3±1
Untreated Bone	1459±130	21±2

The results show that the addition of polymer phase can enhance both Young's modulus and ultimate compressive strength of the scaffolds. However, as discussed in the previous section, vapor bubbles and lack of interfacial adhesion could be problematic and weaken the mechanical properties. Thus, grafting and annealing are required. Table I compares the mechanical performance of natural scaffolds that were treated with different conditions and the corresponding stress-strain curves are shown in Figure 4. Three groups synthesized by different treatments are compared. The first group was only treated with grafting solution and removed from the oven right after complete polymerization (grafting only). The second group remained in the oven for the annealing treatment (grafting and annealing). The third group was directly coated by vapor deposition process followed by annealing without going through the grafting step (annealing only).

Surprisingly, the samples from the first group (grafting only) had the highest Young's modulus and ultimate strength. However, vapor bubbles may remain trapped within the polymer layer without going through the annealing step, as previously shown in Figure 1c. It is observed that samples from the first group had high standard deviation due to the presence of the bubbles. Samples from the third group, on the other hand, showed very limited enhancement due to lack of interfacial adhesion. Therefore, both grafting and annealing treatments should be applied as the standard procedures because these samples show significant enhancement in both Young's modulus and ultimate compressive strength. The mechanical performance of samples from group two (grafting + annealing) is comparable to that of the untreated cancellous bone.

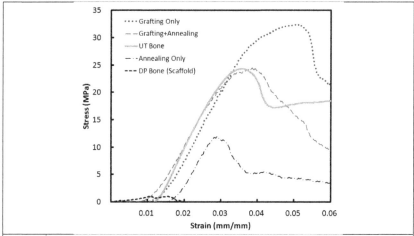

Figure 4. Representative compressive stress-strain curves for the polymer infiltrated DP bone samples prepared by three different treatments (grafting, annealing, and grafting followed by annealing) compared to untreated (UT) bone.

Table II summarizes the compressive mechanical properties of VDP alumina scaffolds synthesized in different conditions. Two controlling factors are the cooling rate and the amount of monomer solution used. The Young's modulus and ultimate compressive strength for polymer infiltrated alumina scaffolds prepared at 2°C/min cooling rate (Figure 5a) and 10°C/min cooling rate (Figure 5b) both increase with increasing the amount of MMA monomer (Figure 5c). However, the scaffolds synthesized at slower cooling rate (2°C/min) have higher Young's modulus and ultimate compressive strength compared with those synthesized at the faster cooling rate (10°C/min). This observation is contradictory to results reported in literatures [6,7] and further experiments need to be conducted. It is possible that the synthesized scaffolds may have some micro-scale damage during sample preparation prior to mechanical tests. The weakening phenomenon may be due to the absence of the degassing process. Without degassing, gas bubbles could be trapped within the liquid slurries and become micro-pores during the solidification process. As the cooling rate increases, gas bubbles do not have enough time to escape from the slurry, which may form cavities or internal cracks in the scaffolds.

Table II Mechanical Properties of PMMA Infiltrated Alumina Scaffolds Synthesized by Vapor Deposition Polymerization

Cooling Rate	10°C/min			2°C/min		
MMA (ml)	0	0.5	1	0	0.5	1
Young's modulus, E (MPa)	272±89	606±485	1394±236	180±51	981±306	2424±739
Ultimate compressive strength, σ (MPa)	4.5±0.7	21±9	48±21	12±4	37±5	78±5

The polymer infiltrated alumina scaffolds prepared by the two different cooling rates demonstrate enhancement in mechanical properties with increasing amount of monomer added. It can be concluded that the mechanical properties of the scaffolds can be tuned by the VDP process. It is possible to synthesize bio-inspired composites that match the mechanical properties of cancellous bone or other hard tissues.

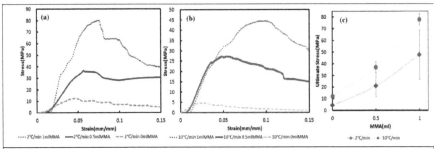

Figure 5. Representative compressive stress-strain curves for the PMMA infiltrated alumina scaffolds freeze cast at (a) 2°C/min (b) 10°C/min cooling rates, respectively. (c) Ultimate compressive strength as a function of the amount of MMA at different cooling rates.

The fracture surface of scaffolds after compression tests is examined under SEM to analyze the failure mechanism. Several failure mechanisms can be observed. Figure 6a shows the fracture surface of DP bone infiltrated by PMMA without grafting treatment. The polymer layer is peeled off from the scaffold after compressive deformation, indicating that there is limited or lack of adhesion between the two phases. In Figure 6b, it can be observed that PMMA being pulled out from the PMMA-infiltrated alumina scaffold during failure creating a rough fracture surface. Under compression, the PMMA-infiltrated alumina scaffolds are typically fractured by the buckling of ceramic walls. The fracture interface is usually located within the PMMA phase but not along the boundary between the polymer/ceramic interface, representing that grafting significantly enhanced the adhesion between the two phases. Figure 6c shows a crack initiates in the PMMA phase (darker region), and deflects around the alumina/PMMA interfaces during propagation, producing a tortuous path. Other toughening mechanisms, such as micro-crack formation in front of the major crack and uncracked ligament bridging can also be observed in Figures 6c and 6d.

CONCLUSIONS
In this study, bio-inspired, polymer-reinforced ceramic scaffolds were synthesized by vapor deposition polymerization and investigated via structural characterization and mechanical testing. Major discoveries and conclusions are summarized as follows:
1. Ceramic scaffolds were prepared by two methods:
   a. Complete deproteinization of bovine cancellous bone
   b. Freeze casting of alumina slurries at two different solidification rates
2. The vapor deposition method was applied to infiltrate polymers (PMMA) into the ceramic scaffolds.
3. Grafting and annealing treatments are required to enhance mechanical properties.
4. The mechanical properties of ceramic scaffolds increase with polymer-infiltration.

5. Bio-inspired composite with tunable microstructure (porosity) and mechanical properties can be synthesized by the combination of freeze casting and vapor deposition polymerization.

6. Toughening mechanisms, such as crack deflection, uncracked ligament bridging and microcrack formation were observed in the polymer-reinforce ceramic scaffolds.

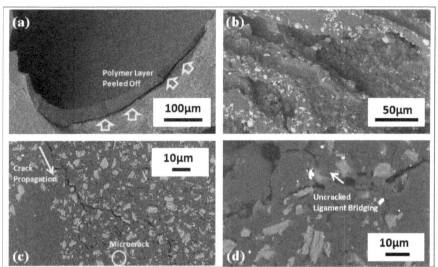

Figure 6. SEM images showing (a) the polymer layer being peeled off at the PMMA/DP bone interface in the PMMA-deposited deproteinized bone without grafting; (b) vapor deposited alumina scaffold with grafting showing rough fracture surface, indicating a ductile failure; (c) crack deflection around the ceramic/polymer interfaces during propagation; (d) microcrack formation and uncracked ligament bridging in PMMA-infiltrated alumina scaffolds.

ACKNOWLEDGEMENTS

We thank Dr. Che-Wei Tsai, Yu-Chen Chan (Department of Material Science & Engineering, NTHU), Jin-Fa Hong (Scientific Instrument Center, NTHU), Dr. Wei-Zhou Chia (National Taiwan University Hospital, Hsinchu branch) and Dr. Bai-Hung Ko (NSRRC) for their technical support and assistance. This work is supported by the National Science Council, Taiwan (NSC100-2218-E-007-016-MY3).

REFERENCES

1. M.A. Meyers, P.Y. Chen, A.Y.M Lin, Y. Seki, Biological materials: Structure and mechanical properties. *Prog in Mater Sci*, **53**, 1-206 (2008)
2. P. Fratzl, R. Weinkamer. Nature's hierarchical materials. *Prog in Mater Sci*, **52**, 1263-1334 (2007)
3. S. Weiner, H.D. Wagner. The material bone: structure-mechanical function relations. *Ann Rev Mater Sci*, **28**, 271-98 (1998)
4. J.D. Currey. Bones: structure and mechanics. Princeton, NJ: Princeton University Press; 2002
5. A. Herzog, R. Klingner, U. Vogt, T. Graule, Wood-derived porous SiC ceramics by sol infiltration and carbothermal reduction, *J. Am. Ceram. Soc.*,**87**, 784-793

6. H. Sieber, C. Hoffmann, A. Kaindl, P. Greil, Biomorphic cellular ceramics, *Adv. Eng. Mater*, **2**, 105-9 (2000)
7. B. Sun, T. Fan, D. Zhang, Porous TiC ceramics derived from wood template, *J. Porous Mater.*, **9**, 275-7 (2002)
8. J. Cao, C.R. Rambo, H. Sieber, J. Preparation of porous Al2O3-Ceramics by biotemplating of wood, *J. Porous Mater.*, 11, 163-72 (2004)
9. P. Sepulveda, J. G. P. Binner, Processing of cellular ceramics by foaming and in situ polymerisation of organic monomers, *J. Eur. Ceram. Soc.*, **19**, 2059-66 (1999)
10. M.E. Launey, E. Munch, D.H. Alsem, H.B. Barth, E. Saiz, A.P. Tomsia, R.O. Ritchie, Designing highly toughened hybrid composites through nature-inspired hierarchical complexity. *Acta Mater.*, **57**, 2919-32 (2009).
11. S. Deville, E. Saiz, R.K. Nalla and A.P. Tomsia, Freezing as a path to build complex composites, *Science*, **311**, 515-18 (2006).
12. E. Munch, M.E. Launey, D.H. Alsem, E. Saiz, A.P. Tomsia, R.O. Ritchie, Tough, bio-inspired hybrid materials. *Science*, **322**, 1516-22 (2008).
13. M.M. Porter, M. Yeh, J. Strawson, T. Goehring, S. Lujan, P. Siripasopsotorn, M.A. Meyers, J. McKittrick, Magnetic freeze casting inspired by nature, *Mat. Sci. Eng. A-Struct.*, **556**, 741-50 (2012).
14. B.W. Riblett, N.L. Francis, M.A. Wheatley, and U.G.K. Wegst, Ice-templated scaffolds with microridged pores direct DRG neurite growth, *Adv. Funct. Mater.*, **22**, 4920–4923 (2012).
15. S. Lee, M. Porter, S. Wasko, G. Lau, P.Y. Chen, E.E. Novitskaya, A.P. Tomsia, A. Almutairi, M.A. Meyers, J. McKittrick, Potential bone replacement materials prepared by two methods, *MRS Proc.,* vol. 1418, (2012) mrsf11-1418-mm06-02 doi:10.1557/opl.2012.671.
16. F.J. Martinez-Vazquez, F.H. Perera, P. Miranda, A. Pajares, F. Guiberteau, Improving the compressive strength of bioceramic robocast scaffolds by polymer infiltration. *Acta Biomater.*, **6**. 4361-68 (2010).
17. X. Miao, W.K. Lim, X. Huang, Y. Chen, Preparation and characterization of interpenetrating phased TCP/HA/PLGA composites. *Mater. Lett.*, **59**, 4000-05 (2005).
18. G. Pezzotti, S.M.F. Asmus, L.P. Ferroni, S. Mikki, In situ polymerization into porous ceramics: a novel route to tough biomimetic materials. *J. Mater. Sci.-Mater. M*, **13**, 783-87 (2002).
19. M.M. Porter, S. Lee, N. Tanadchangsaeng, M.J. Jaremko, J. Yu, M. Meyers, J. McKittrick, Porous hydroxyapatite-polyhydroxybutyrate composites fabricated by a novel method via centrifugation. In XII International Congress and Exposition on Experimental and Applied Mechanics. Costa Mesa of Experimental Mechanics (2012)
20. Y. Elkasabi, H.Y. Chen, J. Lahann, Multipotent polymer coatings based on chemical vapor deposition copolymerization. *Adv. Mater.*, **18**, 1521-26 (2006).
21. P.Y. Chen, J. McKittrick, Compressive mechanical properties of demineralized and deproteinized cancellous bone, *J. Mech. Behav. Biomed.*, **4**, 961-73 (2011).
22. E. Novitskaya, P.Y. Chen, S. Lee, A. Castro-Ceseña, G. Hirata, V.A. Lubarda, J. McKittrick, Anisotropy in the compressive mechanical properties of bovine cortical bone and the mineral and protein constituents, *Acta Biomater.*, **7**, 3170–77 (2011).
23. P.Y. Chen, D. Toroian, P.A. Price, and J. McKittrick, Minerals form a continuum phase in mature cancellous bone, *Calcified Tissue Int.*, **88**, 351-61 (2011).

THE NATURAL PROCESS OF BIOMINERALIZATION AND IN-VITRO REMINERALIZATION OF DENTIN LESIONS

Stefan Habelitz, Tiffany Hsu, Paul Hsiao, Kuniko Saeki, Yung-Ching Chien, Sally J. Marshall, Grayson W. Marshall

Department of Preventive and Restorative Dental Sciences

University of California, San Francisco, USA

ABSTRACT

Biomineralization of collagenous tissues, like bone and dentin, is a complex process which involves the secretion, assembly and organization of matrix molecules which predominantly suppress mineral formation. Only after the processing and modification of collagen and non-collagenous proteins (NCPs) mineral deposition will occur and is, under healthy conditions, highly specific to nucleation site, crystal size and orientation with regards to collagen fibril direction. Incorrect fibril assembly or deficient expression of specific NCPs can severely affect biological mineralization. A model suggesting NCPs predominantly act as nucleators in mineralizing collagenous matrices. Recently a number of studies have shown the ability of polymeric systems to promote mineralization of individual collagen fibrils and matrices leading to the hypothesis that in this in-vitro system charged residues stabilize an amorphous precursor of apatite in solution and facilitate the release of mineralizing ions at site-specific locations on collagen I fibrils, transfuse into the entire volume of the fibrils, to subsequently transform into apatite nanocrystals with their c-axes oriented perpendicular to the fibril long axis. Hence biological mineralization as it occurs in-vivo may follow different mechanisms as observed during mineralization of biological tissues in-vitro. This paper discusses the origin of such differences and presents further evidence on the importance of the type of collagen for facilitating intrafibrillar mineralization.

INTRODUCTION

In-Vivo Biomineralization

While there is a large variety of mineral phases that form in biological organisms including minerals comprised of iron oxides, sulfides, sulfates and carbonates, there may be universal mechanisms that apply to all systems. In this paper, however, we will only discuss the biomineralization principles as they apply to apatite mineral, the calcium phosphate mineral that reinforces collagenous tissue in bones and teeth of vertebrae animals [1-5]. In the craniofacial complex, such tissues are produced by cells that differentiate from mesenchymal stem cells of the neural crest into highly specialized cells that are able to secrete a matrix which is malleable and can be molded into specific shapes as required by the functionality of the tissue or organ it comprises [6].

Precipitation of a mineral by pure thermodynamic driving forces requires that the solution is saturated or in other words that the ion activity product (IP) of the ions in solution is greater than the solubility product, (K_{SP}), which is the equilibrium constant for a solid substance dissolving in an aqueous solution. It represents the level of ion content at which a solute no longer dissolves in the solvent. Hence, for a mineral to be stable and to be able to form a solid particle in an aqueous solution it is a requirement that the Degree of Saturation (DS), defined as the ratio of IP to K_{SP}, to be greater than 1 (see equation 1) [7].

$$DS = IP/K_{SP}, \quad >1 \text{ to form stable nucleus} \qquad (1)$$

Body fluids, like blood or the extracellular fluid in the mineralizing matrix of bone or enamel are significantly saturated with calcium and phosphate ions, e.g. DS > 1, and readily would precipitate calcium phosphate mineral in the absence of an organic phase. Thus a primary task of matrix proteins is to prevent uncontrolled and immediate mineralization when protein is secreted into the extracellular space. The mechanism by which organic molecules inhibit crystallization is attributed to ion binding, predominantly of calcium ions, reducing the free ion concentration, thus lowering DS possibly to values of 1 allowing for an equilibrium between mineral dissolution and precipitation [8]. Acidic residues, often carboxylated, like glutamic acid (Glu) and aspartic acid (Asp) or phosphorylated residues, like serine (Ser), histidine (His) and threonine (Thr) are negatively charged and bind cations to the backbone of the biomacromolecule, thus eliminating these ions from the pool of ions to build a crystal [1, 9]. A knock-out mouse, which lacked the expression of matrix-Gla protein comprised of large numbers of Glu-residues, illustrated the effect that acidic residues have on mineral inhibition in an impressive way, as the entire vascular system in these mice mineralizes causing the mice to die within two months after birth [10].

As a consequence of the high inhibitory potential of organic matrices, mechanisms need to be applied to either reverse or overcome this inhibition and to induce mineralization. It is not entirely understood how evolution dealt with this problem and managed to develop systems that for example allowed collagen fibrils to become internally mineralized, but as a basic principle processing of matrix proteins seems to be required in order to induce mineralization. Most likely all mineralizing systems undergo post-secretory processing by enzymatic activity in order to regulate mineral growth. An evolutionary conserved example is the presence of carbonic anhydrase in corals, shells (including eggshell) that catalyzes the hydration and dehydration of soluble CO_2 and thus directly promotes the formation of carbonate minerals [11]. In apatite mineralizing matrices processing of phosphorylated proteins is known to alter the affinity for proteins to apatite [9]. Prime examples in dentin are dentin sialophosphoprotein (DSPP) and the dentin matrix protein-1 (DMP-1) [12]. Both are initially secreted simultaneously with the collagen matrix, but it is only after their processing that predentin mineralizes to normal levels [13]. In fact, our group has shown that the lack of DSPP as in dentin of patients with dentinogenesis imperfecta type II (DI-II) resulted in a decrease by about 20-30% of mineral, which in return produced extremely poor mechanical properties of dentin with modulus and hardness being only 1/3 of the normal values [14, 15]. This reduced resistance to mechanical load was attributed to the absence of intrafibrillar mineral in dentin collagen of DI-II patients. Studies that compared the potential of NCPs for inducing apatite crystallization by evaluating changes in nucleation lag time or measuring affinity isotherms concluded that the unprocessed and full-length proteins had a lower affinity for calcium phosphate or apatite [16, 17]. Hence a conversion of a mineral inhibitor into a mineral promoter through enzymatic processing is at least in theory possible. Other studies also showed that DPP, the processed form of DSPP, was still acting as an inhibitor when in solution, but once cross-linked to a collagenous matrix it turned into a promoter of apatite crystallization [18]. Other examples include osteopontin in bone, which has cleavage products that have different affinities for apatite when comparing N and C-terminal fragments [16]. Polymer matrices have been designed which upon hydrolysis induce mineralization and may facilitate the formation of bone substitutes by postsurgical mineralization in a defect site [19]. The above observations led to the hypothesis that NCPs in the mineralizing matrices of dentin and bone can act as specific nucleator of apatite and may be located at the gap-zones within collagen fibrils to induce intrafibrillar mineralization [20].

Remineralization of Collagen Fibrils In-Vitro

As the importance of highly charged macromolecules in the biomineralization process became evident numerous studies tested synthetic polymers for their potential to mineralize collagen fibrils. Interestingly such highly charged molecules, like ethylenediaminetetraacetic acid (EDTA) and poly-aspartic acid (poly-Asp) were initially employed to dissolve and remove apatite mineral due to their ability to bind calcium as chelation agents [21]. However the ability to remove calcium from mineral is a function of the concentration of mineralizing ions and the solution pH, with basic pH being more effective. Increasing calcium and phosphate concentration in the presence of these chelation agents, results in apatite mineral being deposited in a controlled manner. In the case of EDTA, a unique organization of fibrous apatite crystals developed when fluoride was added, mimicking an enamel-like material [22]. Initial experiments with poly-Asp and mineralizing solutions were performed in the calcium carbonate system [23], but already established the main concept: In a saturated and metastable solution, the highly charged macromolecules attract calcium and counterions, thus preventing homogenous nucleation in these metastable solutions but stabilizing an amorphous form of the mineral phase. The complex of poly-Asp and amorphous mineral was termed Polymer-Induced Liquid Precursor (PILP) referring to the notion that the inorganic content of the complex did not behave like a solid but had the ability to flow into organic structures suggesting that a substantial amount of water was part of the precursor phase [24]. Mineralization occurred when PILP was brought into contact with organic matrices like type-I collagen fibrils. When PILP was used in combination with calcium phosphate solutions, several studies observed the formation of nanodroplets of an amorphous mineral possibly stabilized by poly-Asp aggregate which specifically bound to the transition of gap to overlap zones on collagen fibrils [25, 26]. Within a period of 24 hours the mineral precursor infiltrated the gap zones and diffused through the fibril. Gradually, over a period of 72 hours, ACP converts into crystalline plate-like apatite with the c-axis oriented along the fibril long axis, comparable to mineral organization in natural tissues [25, 27]. Others proposed that the conversions to oriented apatite required an additional additive [28], but correctly mineralized fibrils were observed without such additives indicating that the type-I collagen molecules have the potential to direct apatite crystallization within fibrils [25, 29]. A recent study obtaining high densities of reconstituted collagen matrices through reverse dialysis was able to replicate the plywood-like arrangement of collagen fibrils as observed in bone and found that intrafibrillar mineral can nucleate in such structures in the absence of the PILP complex but only at extremely high and non-physiological ion concentrations [30]. This supports the hypothesis that collagen type I fibrils by themselves are able to orchestrate the entire mineralization process if assembled correctly and densely packed in the presence of an amorphous calcium phosphate phase. On the other hand, it is in contrast to the observation that plain calcium phosphate solutions do not facilitate intrafibrillar mineralization, but instead promote mineral formation on the surface of the fibrils [26]. We therefore have conducted experiments on type-II collagen fibrils and tested if this type of collagen which does not mineralize in-vivo can be mineralized under in-vitro conditions with and without the PILP system. The findings are described below.

In-vitro (Re-) Mineralization of Demineralized Dentin

Remineralizing dentin lesions is an attractive approach since one of the most prevalent diseases of mankind, dental caries, involves the demineralization of dental structures, like enamel and dentin. While remineralization of enamel, a tissue that consists of approximately 95wt% of crystalline apatite, has been clinical practice for many years [31], reconstituting dentin by a chemical process is a more challenging endeavor due to the presence of an organic matrix which remains and becomes the scaffold that needs to be remineralized after a caries attack. Initial studies showed that remineralization applications with and without stabilizing additives will be able to recover mineral content in artificial dentin lesions at shallow depths [32-34]. A constant composition approach, which continuously replenished the mineralizing solution with calcium and phosphate ions, was successful in remineralizing dentin and recovered properties of the tissue to about 60% of normal for lesions that were about 30 μm in depth [32]. Deeper in-vitro lesions, designed to penetrate through enamel into dentin, were also remineralized by reincorporating mineral from the bottom of the lesion upwards to the surface when in contact with calcium phosphate solutions over extended time periods, but an analysis of concomitant mechanical recovery was not performed [35]. A systematic evaluation of nanomechanical properties across lesion profiles showed that full recovery of the mineral content after PILP treatment is not necessarily associated with full recovery of the mechanical behavior of the remineralized dentin [29]. Apatite mineral may form in the collagenous meshwork of dentin but without association to the organic matrix or alternatively mineral may have precipitated in the lumen of dentin tubules, thus indicating a successful remineralization treatment as mineral content increases when analyzed by X-ray tomography or dental radiographs. The properties of such a re-mineralized structure are however far from the values obtained on the sound tissue, as the mineral portion is not actively integrated with the organic matrix. In these deeper lesions we found that only a narrow portion, the transition zone adjacent to the unaffected and normal dentin actually recovered functionally and fully restored the elastic modulus and hardness of dentin [29].

In this study we investigated in more detail if saturated calcium phosphate solutions alone, including the constant composition approach, were sufficient to functionally remineralize deeper lesions in dentin or if acidic polymers, e.g. poly-Asp, were required to recover properties across the entire depth of an artificial lesion. In addition we perform tests on type-II collagen fibrils to investigate if collagen fibrils that naturally do not mineralize will also be able to block intrafibrillar mineralization in-vitro, supporting the idea that collagen type I structure is inductive of apatite mineralization by itself.

MATERIALS AND METHODS

Dentin blocks from human third molars were prepared as previously described and exposed to undersaturated calcium phosphate solutions at pH 5.0 for a total of 66 hours [29]. Subsequently samples were rinsed with deionized water and immersed into 3 different remineralization solutions, all containing 4.5 mM $CaCl_2$ and 2.5 mM KH_2PO_4 with and without additives: A) CaP only, no additives; B) CC-CaP: Constant composition approach using a titrator and dosimat which titrate Ca^{2+} and PO_4^{3-} ions from two separate containers into a reaction vessel triggered by a pH drop, as previously described [32, 36]. The pH was adjusted to 7.4 using KOH, no other additives used; C) PILP: Calcium phosphate solution with addition of 100 μg/ml poly-Asp (Poly(L-aspartic acid sodium salt), 27-kDa, Alamanda Polymers, Inc) was used. All solutions had an initial volume of 200 ml with three samples per solution. Specimens were removed after 14 days, rinsed with deionized water and cut perpendicularly through the lesion for analysis of their cross sections by nanoindentation while fully immersed in water. Indentations were performed with the Hysitron-Triboscope system on a Digital Instrument Multi-Mode AFM (Nanoscope III). A Berkovich indenter with long shaft for fluid immersion was calibrated using silica glass. Indentations were made along a line at intervals of 3 to 5 μm from the embedding material through the artificial lesion into the normal dentin. At least three lines of indentations were made per specimen and data was averaged over all samples studied per group (n =3) using statistical methods as described earlier [29].

Commercially available collagen type II from chicken sternal cartilage (Sigma–Aldrich, USA) was used to self-assemble into fibrils and to study their potential for remineralization following a published protocol [37]. Briefly, lyophilized collagen powder was dissolved in acetic acid (pH 3.0, 4°C) overnight, then self-assembly was induced by pH change using a buffer exchange by dialysis (dialysis buffer: 50mM L-glycine, 200mM KCl, 0.5mg/ml collagen; adjusted to pH 9.3 with 1M NaOH, 37°C). Equal volumes of calcium (9mM $CaCl_2$) and phosphate (4.2mM K_2HPO_4) stock solutions were added to dehydrated fibrils on a glass cover slip glued to an AFM slide for reaction times 4, 6, 12, 24, 48 hours. To prevent evaporation, AFM disks were kept in a plastic dish with wet filter paper and Parafilm to seal the edges. Stock solutions were stored in Tris-buffer solutions (pH 7.4, 4°C). Calcium solution contained 0.2 mg/ml poly-Asp. Control reactions contained no poly-Asp. Atomic Force Microscopy (AFM, Nanoscope III, Digital Instruments, Santa Barbara, CA) in dry tapping mode with a silicon tip was used to measure fibril height (h) and width (w) (n=10-20 fibrils/sample). Collagen fibril diameter contracts in its vertical axis during dehydration while expanding in the horizontal axis, thus becoming oval shaped when dehydrated. The collapse (C) of a single fibril was specified as

$$C = (1- (h/w)) \times 100\% \qquad (2)$$

As described in a previous study on dentin, using an attached optical microscope we were able to image the exact same location and the same fibril before and after the mineralization treatment[38]. The fibril collapse was measured before (C_{before}) and after treatment (C_{after}) with and without PILP at different time points and compared by student t-test. The difference of $C_{(after)}$ – $C_{(before)}$ was then used as an indicator for successful mineralization treatment as we assume that a reduced the degree of collapse is associated with mineral being introduced into the fibril, see Table 1.

Fourier-transformed spectral analysis was used to also determine the periodicity of individual fibrils [39]. Transmission electron microscopy (TEM) with nickel grids was used to observe collagen fibrils at mineralization times for up to 7 days. A drop of collagen buffer was placed on a nickel grid, lightly blotted with filter paper, and then placed upside down into a sealed microcentrifuge tube of calcium and phosphate solution.

RESULTS AND DISCUSSION

The demineralization process using acetic acid buffers at pH 5.0 for 66 hours resulted in a demineralized zone of about 160 μm, which was the depth at which modulus values obtained were close to normal dentin values (see yellow circles in Fig. 1). The artificial lesion is further characterized by a zone of gradient properties. The modulus gradually decreases from normal dentin values around 18 GPa to a constant and flat zone of low stiffness with modulus values below 0.5 GPa. The gradient zone is about 100 μm wide and reaches from 160 μm depth to about 60 μm. The outer zone is extremely soft and most likely consists of only of hydrated organic matrix of dentin with little residual mineral, while the gradient zone is believed to contain increasing amounts of extra-and intra-fibrillar mineral residue which have not fully dissolved during the demineralization process. Crystals may also regrow during demineralization as the undersaturation of the solution is very small, $DS_{HAP} = 0.01$ and locally the solution can become saturated and induce recrystallization.

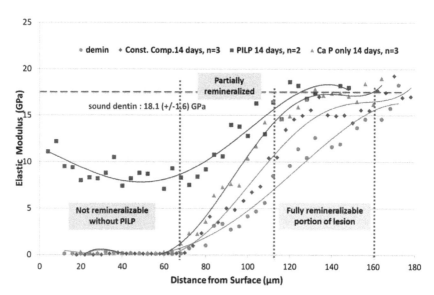

Figure 1: Elastic Modulus plot across artificial dentin lesions comparing different remineralization methods at 14 days of treatment as a function of depth of the lesion measured as distance from the surface: (circles) demineralized control, (diamonds) remin with CC-CaP, (triangles) remineralized with CaP, and (squares) remineralized using PILP.

When such artificial lesions are exposed for 14 days to mineralization solutions containing calcium and phosphate only, remineralization of the lesion was observed, but only in a narrow zone adjacent to the normal dentin. As shown in Figure 1, about 50 µm of the tissue fully recovered, reaching an elastic modulus of 15 GPa and higher to a lesion depth of about 110 µm. This remineralizable zone is followed by a gradient zone which transitions into the fully demineralized zone at about the same depth (65 µm) as the original demineralized dentin which was not remineralized. Properties in this partially mineralized zone have only slightly improved compared to lesions that were not treated with mineralizing solutions. The outer zone from 65 µm depth to the surface of the lesion did not improve in mechanical response remaining flat at values of 0.5 GPa and lower that is approxiamtely the lower limit to be accurately determined by the Triboscope nanoindentation system used. Hence it appeared that mineral was not deposited in the outer zone in a functional way.

Identical calcium phosphate solutions were prepared but in addition a constant composition titration system was used to maintain pH and mineralizing ion content over the 14 days of treatment. A total volume of up to 22 ml was added during this period and pH fluctuation was minimal between pH 7.39 and 7.41. Using CC-CaP, it was expected that mineralization is enhanced allowing for continued growth of mineral during the entire period of treatment since DS remains approximately constant. However, we did not observe significant differences between CaP-only and CC-CaP as both showed an almost identical profile of moduli along cross sections through the lesion (Fig. 1). This may be due to the large volume of mineralizing solution used which does supply enough calcium and phosphate ions and most likely prevents the solutions from significant changes in DS or from becoming undersaturated. We also noticed a tendency of specimens that were remineralized with CC-CaP to form a superficial mineral layer. Hence some of the mineral ions added by the titration system were consumed by mineral formation at the surface and did not help to restore the lesion.

In contrast to the previous two approaches, remineralization using solutions with the same calcium and phosphate content but with addition of poly-Asp acid resulted in a significant increase in the elastic modulus in all zones of an artificial dentin lesion (Fig. 1). At the 14 day time point investigated here, the inner lesion was fully restored by PILP treatments, very similar to CaP and CC-CaP treatments. The zone that is partially remineralized had improved properties when compared to CaP and CC-CaP. The most significant change, however, became evident at the outer zone. PILP treatments resulted in a substantial increase of modulus achieving values close to 10 GPa in this zone that was below 0.5 GPa before the remineralization treatment. As shown in a previous study, the presence of poly-Asp during remineralization had a strong impact on the properties of the dentin matrix, which may be related to the ability of PILP to induce intrafibrillar mineralization of the collagen fibrils [29]. A more detailed analysis by TEM and other techniques is warranted to identify the origin of such remarkable recovery. It should be noted, however, that only about 50 to 60% of the original elastic properties of normal dentin were recovered, suggesting that there might be other structural differences between sound dentin and dentin remineralized by PILP such as differences in mineral size, total extrafibrillar and intrafibrillar mineral content.   Structural and biochemical changes may also have been introduced during the demineralization process as matrix proteins become activated and mobile and alter crosslinking of collagen fibrils or remove certain NCPs that faciliate a bond between organic matrix and inorganic crystals [40]. The type-I collagen fibrils may be able to induce conversion of ACP into oriented apatite platelets by itself, but additional factors might be relevant when attempting to reconstitute bulk properties of a mineralized collagenous tissue.

In a second set of experiments we tested if the ability of type-I fibrils to mineralize is unique and intrinsic to its structure by testing if type-II fibrils, which in natural tissue do not mineralize, can be mineralized using the PILP system. We used AFM and TEM analysis to detect mineral formation in the fibrils. When analyzed over a period between 4 and 24 hours we observed that type-II fibrils actually had a smaller height to width ratio (h/w) when PILP or CaP solutions were applied compared to before the treatment (Table 1). The fibril collapse appeared stronger after te mineralization treatment, but was actually related to an increase in fibril width. The fibrils became wider due to mineral buildt up at the surface of the fibril, while the height only slightly increased. Analysis of AFM images also suggested that mineralization has occured on the surface of the fibrils instead of internally (Fig. 2 a-c). At 48 hours of treatment the difference in fibril collapse was smaller for both treatment groups. Interestingly the only treatments that resulted in an overall reduction of fibril collapse was the 48 hour PILP treatment. This could indicate that at this point calcium phosphate has entered the fibril and is now supporting the integrity of the fibril and reducing a collapse upon dehydration. TEM of remineralized type-II fibrils showed that mineral form in association with the type-II fibrils when treated by PILP. A continuous layer of an electron-dense material can be observed in Fig. 2d and

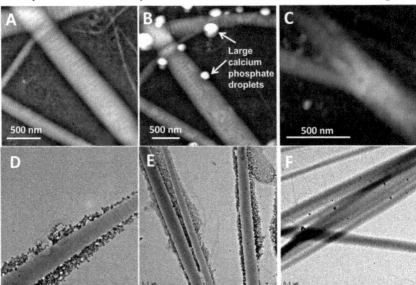

Figure 2: Micrographs of type II collagen fibrils at different stages of mineralization treatments using the PILP process: a-c) Atomic Force Microscopy images of type II fibrils after assembly (a); after PILP treatment for 5 h (b); after PILP treatment for 72 h (c); TEM images of unstained type II fibril treated for 5 days using PILP (d and e) and using CaP-only f).

e; such layer was absent when remineralization was performed using CaP-only (Fig. 2f). TEM analysis also showed that mineral did not form within type-II fibrils (Fig. 2d, e) suggesting that either the PILP method, at the conditions used, is not suitable for mineralization of this type of fibril or that indeed an intrinsic mechanism exists that prevents type-II fibrils from being mineralized internally. Nevertheless, TEM analysis of collagen II fibrils treated with PILP had a significantly increased contrast revealing the characteristic periodicity pattern of collagen fibrils. Such pattern can usually only be resolved with thungsten or uranyl staining or when collagen fibril was adequately mineralized in the gap zones. We did not observe the presence of mineral or oriented calcium phosphate reflexes using electron diffraction and therefore exculde the presence of crystlline apatite in the collagen II fibrils at the time points studied. The improved contrast in the type II fibril patterning may indicate that an amorphous phase is present in the gap zones of the fibrils but has not mineralized. Inhibition to intrafibrillar mineralization could be intrinsic to type II collagen fibrils. This further supports the hypothesis that the primary structure of type-I collagen fibrils or other mineralizing collagens might be instructive in directing a phase transformation from ACP to crystalline HAP and acts as a template for oriented crystallization of apatite platelets.

Table 1: AFM data on type-II collagen fibrils remineralzied by CaP-only and PILP for up to 48 hours.

Mineral. Time	Condition	# of Fibrils	$C_{(after)} - C_{(before)}$	P-value $C_{(after)} / C_{(before)}$
4 hrs	With pAsp	20	- 4.06 ± 6.80	1.2E-06
	No pAsp	20	- 4.33 ± 3.65	4.1E-09
12 hrs	With pAsp	10	- 7.30 ± 17.99	4.1E-04
	No pAsp	10	- 8.20 ± 7.21	4.8E-06
24 hrs	With pAsp	20	- 4.05 ± 5.11	1.6E-07
	No pAsp	20	- 4.79 ± 3.28	3.4E-10
48 hrs	With pAsp	15	**+0.07 ± 1.78**	**0.836**
	No pAsp	15	- 1.50 ± 1.04	5.44E-05

CONCLUSIONS

Remineralization of dentin tissue is a complex process. The re-introduction of apatite mineral into demineralized dentin collagen fibrils is possible, but properties of the remineralized tissue strongly depend on the remineralization procedure as well as on the zone of the artificial lesion. It appears that any treatment with any saturated calcium phosphate solution can functionally remineralize a portion of dentin which still contains remnants of apatite mineral with residual crystallites acting as nucleation sites . Once the fibril has lost all of its mineral, participation of a liquid precursor phase of apatite becomes essential to the mineralizing process. Poly-Asp was found to succesfully reintroduce apatite mineral platelets into dentin collagen fibrils. Type-I collagen is able to induce the oriented crystallization of apatite platelets in the gap zone of fibrils. An ability that may be related to the primary structure of type-I collagen and other collagen types that mineralize, as PILP treatments of type-II fibrils did not succeed in introducing calcium phosphate mineral. Similarly it has been shown that type-I collagen fibrils from tendon of the rat tail does not mineralize via PILP treatments without pretreatment and removal of NCPs [24]. Remineralization of dentin by PILP is accompanied by a substantial increase in the properties. The treatment recovers stiffness in the outer zone which completely lacked apatite mineral, but only to about 50% of the values reported for sound dentin, indcating structural or biochemical differences in the remineralized tissue.

There is great potential for the use of poly-Asp for remineralization of dentin carious lesion as the PILP method is able to reintroduce apatite crystallites into collagen fibrils. While this study provides further evidence that intrafibrillar mineralziation contributes a large portion to the mechanical resitance to deformation of dentin it also shows that other factors, including parameters of the demineralziation process, will affect the success of treatments aimed to reconstitute the original structure and properties of dentin.

ACKNOWLEDGEMENT

We appreciate the support by Dr. Laurie Gower and Dr. Taili Thula-Mata (U. Florida, Gainesville) in performing the PILP experiments. We also would like to thank Grace Nonomura and Anora Burwell (UCSF) for specimen preparation. Financial support by NIH/NIDCR RO1-DE016849 and UCSF-CTSI-SOS grant #166.

REFERENCES

[1] Addadi, L. and S. Weiner, *Interactions between acidic proteins and crystals: stereochemical requirements in biomineralization.* Proc Natl Acad Sci U S A, 1985. **82**(12): p. 4110-4.

[2] Mann, S. and S. Weiner, *Biomineralization: structural questions at all length scales.* J Struct Biol, 1999. **126**(3): p. 179-81.

[3] Williams, R.J., *Biomineralization: iron and the origin of life.* Nature, 1990. **343**(6255): p. 213-4.

[4] Gower, L.B., *Biomimetic Model Systems for Investigating the Amorphous Precursor Pathway and Its Role in Biomineralization.* Chemical Reviews, 2008. **108**(11): p. 4551-4627.

[5] Lowenstam, H.A., *Minerals formed by organisms.* Science, 1981. **211**(4487): p. 1126-31.

[6] Thesleff, I. and P. Nieminen, *Tooth morphogenesis and cell differentiation.* Curr Opin Cell Biol, 1996. **8**(6): p. 844-50.

[7] Orme, C. and J. Giocondi, *Model Systems for Formation and Dissolution of Calcium Phosphate Minerals*, in *Biomineralization*, V. Bauerlein, Editor 2008, VCH.

[8] Hauschka, P.V. and F.H. Wians, *Osteocalcin-Hydroxyapatite Interaction in the Extracellular Organic Matrix of Bone.* Anatomical Record, 1989. **224**(2): p. 180-188.

[9] Glimcher, M.J., *Mechanism of calcification: role of collagen fibrils and collagen-phosphoprotein complexes in vitro and in vivo.* Anat Rec, 1989. **224**(2): p. 139-53.

[10] Luo, G.B., P. Ducy, M.D. McKee, G.J. Pinero, E. Loyer, R.R. Behringer, and G. Karsenty, *Spontaneous calcification of arteries and cartilage in mice lacking matrix GLA protein.* Nature, 1997. **386**(6620): p. 78-81.

[11] Marie, B., G. Luquet, L. Bedouet, C. Milet, N. Guichard, D. Medakovic, and F. Marin, *Nacre calcification in the freshwater mussel Unio pictorum: carbonic anhydrase activity and purification of a 95 kDa calcium-binding glycoprotein.* Chembiochem : a European journal of chemical biology, 2008. **9**(15): p. 2515-23.

[12] Qin, C., O. Baba, and W.T. Butler, *Post-translational modifications of sibling proteins and their roles in osteogenesis and dentinogenesis.* Critical reviews in oral biology and medicine : an official publication of the American Association of Oral Biologists, 2004. **15**(3): p. 126-36.

[13] Butler, W.T., J.C. Brunn, and C. Qin, *Dentin extracellular matrix (ECM) proteins: comparison to bone ECM and contribution to dynamics of dentinogenesis.* Connective tissue research, 2003. **44 Suppl 1**: p. 171-8.

[14] Kinney, J.H., S. Habelitz, S.J. Marshall, and G.W. Marshall, *The importance of intrafibrillar mineralization of collagen on the mechanical properties of dentin.* J Dent Res, 2003. **82**(12): p. 957-61.

[15] Kinney, J.H., J.A. Pople, C.H. Driessen, T.M. Breunig, G.W. Marshall, and S.J. Marshall, *Intrafibrillar mineral may be absent in dentinogenesis imperfecta type II (DI-II).* J Dent Res, 2001. **80**(6): p. 1555-9.

[16] Boskey, A.L., B. Christensen, H. Taleb, and E.S. Sorensen, *Post-translational modification of osteopontin: effects on in vitro hydroxyapatite formation and growth.* Biochemical and biophysical research communications, 2012. **419**(2): p. 333-8.

[17.]Hunter, G.K., H.A. Curtis, M.D. Grynpas, J.P. Simmer, and A.G. Fincham, *Effects of recombinant amelogenin on hydroxyapatite formation in vitro.* Calcif. Tissue Int., 1999. **65**(3): p. 226-31.

[18.]Saito, T., M. Yamauchi, Y. Abiko, K. Matsuda, and M.A. Crenshaw, *In vitro apatite induction by phosphophoryn immobilized on modified collagen fibrils.* J. Bone Miner. Res., 2000. **15**(8): p. 1615-9.

[19.]Murphy, W.L. and D.J. Mooney, *Bioinspired growth of crystalline carbonate apatite on biodegradable polymer substrata.* Journal of the American Chemical Society, 2002. **124**(9): p. 1910-1917.

[20.]Landis, W.J., *An overview of vertebrate mineralization with emphasis on collagen-mineral interaction.* Gravitational and space biology bulletin : publication of the American Society for Gravitational and Space Biology, 1999. **12**(2): p. 15-26.

[21.]Littlejohn, F., A.E. Saez, and C.S. Grant, *Use of Sodium Polyaspartate for the Removal of Hydroxyapatite/Brushite Deposits from Stainless Steel Tubing.* Ind. Eng. Chem. Res., 1998. **37**: p. 2691-2700.

[22.]Chen, H., B.H. Clarkson, K. Sun, and J.F. Mansfield, *Self-assembly of synthetic hydroxyapatite nanorods into an enamel prism-like structure.* Journal of colloid and interface science, 2005. **288**(1): p. 97-103.

[23.]Olszta, M.J., E.P. Douglas, and L.B. Gower, *Scanning electron microscopic analysis of the mineralization of type I collagen via a polymer-induced liquid-precursor (PILP) process.* Calcif Tissue Int, 2003. **72**(5): p. 583-91.

[24.]Olszta, M.J., X.G. Cheng, S.S. Jee, R. Kumar, Y.Y. Kim, M.J. Kaufman, E.P. Douglas, and L.B. Gower, *Bone structure and formation: A new perspective.* Materials Science & Engineering R-Reports, 2007. **58**(3-5): p. 77-116.

[25.]Nudelman, F., K. Pieterse, A. George, P.H. Bomans, H. Friedrich, L.J. Brylka, P.A. Hilbers, G. de With, and N.A. Sommerdijk, *The role of collagen in bone apatite formation in the presence of hydroxyapatite nucleation inhibitors.* Nature materials, 2010. **9**(12): p. 1004-9.

[26.]Liu, Y., Y.K. Kim, L. Dai, N. Li, S.O. Khan, D.H. Pashley, and F.R. Tay, *Hierarchical and non-hierarchical mineralisation of collagen.* Biomaterials, 2011. **32**(5): p. 1291-300.

[27.]Thula, T.T., D.E. Rodriguez, M.H. Lee, L. Pendi, J. Podschun, and L.B. Gower, *In vitro mineralization of dense collagen substrates: A biomimetic approach toward the development of bone-graft materials.* Acta Biomaterialia, 2011. **7**(8): p. 3158-3169.

[28.]Kim, Y.K., L.S. Gu, T.E. Bryan, J.R. Kim, L. Chen, Y. Liu, J.C. Yoon, L. Breschi, D.H. Pashley, and F.R. Tay, *Mineralisation of reconstituted collagen using polyvinylphosphonic acid/polyacrylic acid templating matrix protein analogues in the presence of calcium, phosphate and hydroxyl ions.* Biomaterials, 2010. **31**(25): p. 6618-27.

[29.]Burwell, A., T. Thula-Mata, L. Gower, S. Habelitz, M. Kurylo, S.P. Ho, Y.-C. Chien, J. Cheng, N.F. Cheng, S.A. Gansky, S.J. Marshall, and G.W. Marshall, *Functional Remineralization using Polymer-Induced Liquid-Precursor Process, .* PloS one, 2012. **7**(6): p. e38852.

[30.]Wang, Y., T. Azais, M. Robin, A. Vallee, C. Catania, P. Legriel, G. Pehau-Arnaudet, F. Babonneau, M.M. Giraud-Guille, and N. Nassif, *The predominant role of collagen in the nucleation, growth, structure and orientation of bone apatite.* Nature materials, 2012. **11**(8): p. 724-33.

[31.]ten Cate, J.M. and J.D. Featherstone, *Mechanistic aspects of the interactions between fluoride and dental enamel.* Critical reviews in oral biology and medicine : an official publication of the American Association of Oral Biologists, 1991. **2**(3): p. 283-96.

[32] Bertassoni, L.E., S. Habelitz, S.J. Marshall, and G.W. Marshall, *Mechanical recovery of dentin following remineralization in vitro--an indentation study.* Journal of biomechanics, 2011. **44**(1): p. 176-81.

[33] Tay, F.R. and D.H. Pashley, *Guided tissue remineralisation of partially demineralised human dentine.* Biomaterials, 2008. **29**(8): p. 1127-37.

[34] Tay, F.R. and D.H. Pashley, *Biomimetic remineralization of resin-bonded acid-etched dentin.* J Dent Res, 2009. **88**(8): p. 719-24.

[35] ten Cate, J.M., *Remineralization of caries lesions extending into dentin.* Journal of dental research, 2001. **80**(5): p. 1407-11.

[36] Uskokovic, V., M.K. Kim, W. Li, and S. Habelitz, *Enzymatic processing of amelogenin during continuous crystallization of apatite.* Journal of Materials Research, 2008. **23**(12): p. 3184-3195.

[37] Franz, C.M. and D.J. Muller, *Studying collagen self-assembly by time-lapse high-resolution atomic force microscopy.* Methods in molecular biology, 2011. **736**: p. 97-107.

[38] Marshall, G.W., N. Yucel, M. Balooch, J.H. Kinney, S. Habelitz, and S.J. Marshall, *Sodium hypochlorite alterations of dentin and dentin collagen.* Surface Science, 2001. **491**(3): p. 444-455.

[39] Habelitz, S., M. Balooch, S. Marshall, G. Balooch, and G. Marshall, *In situ atomic force microscopy of partially demineralized human dentin collagen fibrils.* J. Struct. Biol., 2002. **138**(3): p. 227.

[40] Tay, F.R. and D.H. Pashley, *Dental adhesives of the future.* J Adhes Dent, 2002. **4**(2): p. 91-103.

SYNTHESIS OF HIGHLY BRANCHED ZINC OXIDE NANOWIRES

Wenting Hou[†], Louis Lancaster[†], Dongsheng Li[†‡], Ana Bowlus[†], David Kisailus[†]

[†]Department of Chemical and Environmental Engineering, University of California, Riverside, California 92521, United States

[‡]Materials Science Division, Lawrence Berkeley National Laboratory, Berkeley, California 94720, United States

ABSTRACT

Biological mineralizing processes demonstrate how Nature can produce elegant structures through controlled organic-mineral interactions. These organics are often used to control shape, size and orientation of mineral. Based on inspiration from Nature, which often use organic-mineral or organic-ion interactions to control crystallization processes, we utilize an organic agent, Ethylenediamine (EDA), as a mineralizer to inhibit rapid hydrolysis and condensation of ZnO, and thus control crystal growth behavior. Through adjusting various parameters, such as precursor concentration as well as the molar ratio of the inorganic precursor and organic ligands, we were able to produce highly branched ZnO nanostructures, which is very promising for dye-sensitized solar cells (DSSCs). We also investigate the mechanism of this branching event.

INTRODUCTION

Due to the increasing demand for energy and the environmental pressure caused by fossil fuels, significant progress has been made to develop alternative renewable energy[1]. Solar energy, as a carbon-neutral process, is considered the ultimate solution to replace fossil fuel and meet the environmental challenge. Dye Sensitized Solar Cells (DSSCs) is a promising device at low cost, which is able to compete the available commercialized high cost solar cells based on Silicon and compound semiconductors[2-4]. However, further increase of the DSSC conversion efficiency[4, 5, 6] (11.4%) has been limited due to electron recombination in bulks and at the interfaces of the materials as well as limited coverage of dye molecules on the surfaces. A material with high electron mobility would reduce the recombination rate and an enlarging surface area would be available for the adsorption of more dye molecules. ZnO, a II-VI compound semiconductor with a direct bandgap of 3.37 eV, a high exciton binding energy (60meV)[7, 8], and an electron mobility as high as 2000[7, 9], would be a promising candidate compared to the widely used TiO_2[9,10]. ZnO has been identified as a promising "future material"[11]. In fact, ZnO has significant potential to be used in technologies other than photovoltaic devices including sensors, ultra-violet laser diodes and photocatalysts[8, 11, 12]. To improve the conversion efficiency of DSSCs, numerous efforts have been made to synthesize ZnO with different structures such as nanoparticles[13], nanowires[14, 15], nanotubes[16] and thin films[17, 18]. Morphology, crystallinity and orientation of ZnO crystals should have a strong effect on its resulting properties[19]. For example, nanoparticles networks with disordered pore structures have been observed to exhibit slow electron transport kinetics due to formation of electron traps at the contacts between nanoparticles. Conversely, aligned single crystalline nanowires are promising for fast electron transportation due to lack of grain boundaries acting as traps[14, 20] and thus, the replacement of nanoparticles with nanowires have shown an enhanced electrical conductivity[14, 15, 20]. Although the nanowires show better electron transport kinetics, they have a reduced surface area, which limits the amount of sensitizer (e.g., dye molecules in DSSCs) that can be absorbed. Thus, in order to enhance DSSC efficiency, it is necessary to produce ZnO nanostructures with both good electron transport and high surface area[21].

ZnO can be synthesized by a variety of methods including physical[22] or chemical vapor deposition[23, 24], which require extreme conditions like high temperature or high vacuum to achieve the desired size and phase of the materials. Solution routes such as chemical bath deposition[25, 26], sol-gel synthesis[27] and hydro/solvo-thermal methods[28] utilize relatively mild temperatures and pressures and afford low costs, yet these often lack precise control over size and morphology of the resulting crystals. Over billions of years, Nature has evolved to produce elegant structures that serve highly specified functions. This is exemplified in biomineralized organisms, which are able to control the nucleation location, growth orientation, phase of the mineral, as well as the size and shape of the crystallites[29, 30]. These organisms can do this by utilizing organics to template the nucleation and growth of inorganic materials with incredible precision and fidelity all under mild conditions (room temperature and near-neutral pH). Examples of these biominerals include calcite[31-33], iron oxide[34] and silica[35], all of which are grown in the presence of organic scaffolds and soluble organic compounds. During the growth, minerals or ions that constitute the minerals interact with these organics, and subsequently grow in a directed fashion, yielding different morphologies. Based on inspiration from the formation mechanisms of these biological minerals, we utilize functional organic moieties to direct the growth of ZnO under hydrothermal conditions. We investigate how modified solution parameters affect the growth and branching of nanocrystalline structures.

MATERIALS AND METHODS

ZnO nanostructures were synthesized under hydro-solvothermal conditions in 23mL Teflon liners. Concentrated precursor solutions of zinc were prepared by dissolving Zinc nitrate hexahydrate ($Zn(NO_3)_2 \cdot 6H_2O$, 98%, ACROS) and sodium hydroxide (NaOH) in Milli-Q water to form an aqueous precursor solution (molar ratio $[Zn^{2+}]:[OH]=1:20$). Mixtures of water/ethanol were used as co-solvents and ethylenediamine (EDA) was added as a mineralizer. Specific quantities of the precursor solution were added to water/ethanol co-solvents to make different concentration zinc nitrate solutions. Subsequently, different amounts of EDA solutions were added into the solutions. The final solution precursor mixtures (i.e., $Zn(NO_3)_2$, EDA, water and ethanol) were placed into 23mL Teflon-lined autoclaves (Parr Instruments, Moline, IL), sealed and placed in convection ovens at 180°C for different durations (10 minutes - 20 hours). After the reactions were complete, reactors were removed and subsequently cooled under ambient conditions. The resulting products were then washed three times with DI water and twice with ethanol, with product suspensions being sonicated (Branson 2510, Danbury, CT) between washes to remove any unreacted precursor and reaction by-products. Samples were then dried in air at room temperature.

The resulting ZnO products were characterized using X-Ray Diffraction (XRD, Philips X'Pert), using Cu Kα radiation ($\lambda = 1.5405$ Å) for phase analysis. Structural analyses were conducted using Scanning Electron Microscopy (SEM, Philips FEI XL30) at 10kV accelerating voltage, Transmission Electron Microscopy (TEM, CM300) operated at 300kV. Chemical analysis was conducted by Fourier Transform Infrared Spectroscopy (FTIR, Bruker Equinox 55, 4000 cm^{-1} to 370 cm^{-1}, step size 1 cm^{-1}). Samples for FTIR were prepared by grinding dried potassium bromide (KBr) with 1% ZnO sample in a mortar and pestle, and drying at 60°C for 4 hrs. 100mg of the KBr and ZnO powders were subsequently pressed into pellets for FTIR analyses.

RESULTS AND DISCUSSION

Effect of Zn Concentration

Figure 1. SEM micrographs and XRD patterns of ZnO nanostructures synthesized at 180°C, pH=14 and [Zn]: [EDA]=1:1, at different [Zn]: (a) [Zn]=5mM, (b) [Zn]=10mM, (c) [Zn]=25mM, and (d) [Zn]=50mM. XRD (right) indicates a decrease in crystallite size with increasing [Zn].

The size and morphology of ZnO nanoparticles have a great influence on its optoelectronic properties, which is likely to affect performance in DSSCs[9]. Here, different ZnO nanostructures have been prepared using precursor solutions with different zinc concentrations with EDA as a mineralizer. XRD (Figure 1) confirms that all the samples are in the form of crystalline wurtzite ZnO (JCPDS 36-1415). No additional phases (such as zinc hydroxide) were observed. The crystallinity of the samples decreased from 50nm (at 5mM) to 20nm (at 50mM) zinc precursor concentration (as observed by the increase in the full width at half maximum (FWHM) of the (100) ZnO peak). From the XRD observation, it is clear that the relative intensity of ZnO (001) peak increased slightly in the [Zn]=50 mM sample compared to the samples at lower zinc precursor concentrations, which might be induced by the formation of the branches on the prismatic surfaces of ZnO core rods. SEM micrographs of these structures (Figure 1), revealed obvious differences between the ZnO structures produced using different precursor concentrations. At low concentrations of zinc precursor (5mM), long (>3μm) and smooth ZnO rods formed with diameters between 50-200 nm. With increasing precursor concentration (10mM, 25mM), the length and aspect ratio of ZnO rods decreases from 3μm and 25 to 500nm and 4, respectively. When the concentration of zinc precursor is increased to 50mM, highly branched ZnO rods formed (compared to nanoparticles or smooth rods at low concentrations). These ZnO branched structures consist of long (5-10 μm) hexagonal rods with highly condensed, oriented nanobranches growing from its six prismatic (10-10) faces. At low nutrient concentrations (i.e., low concentration of hydrolyzable species), crystal growth is favorable whereas at higher concentrations, nucleation would dominate. The rate of nucleation grows rapidly with increasing precursor concentration[36]. Therefore, at high concentrations (50mM), nucleation occurs in a very short period of time, producing a large number of nuclei, which would not only decrease the crystalline size of ZnO, but also affect the manner in which the structures grow. Li et al. have demonstrated that high concentrations of titania precursor lead to spherulitic nanostructures, consisting of a core nanowire with multiple branches that increase in density with increasing Ti precursor concentration and time[37]. Thus, in our system, it is likely that at higher concentrations of Zn precursor, more nuclei were produced, increasing the probability of a branching event via twinning[38].

Branched Structure Analysis

SEM and TEM analyses were used to further investigate the characteristics of the highly branched structures (Figure 2). SEM observations (Figure 2a, d) showed that the core of the ZnO branched nanostructures had diameters between 1μm-2μm, and the secondary rods (i.e.,

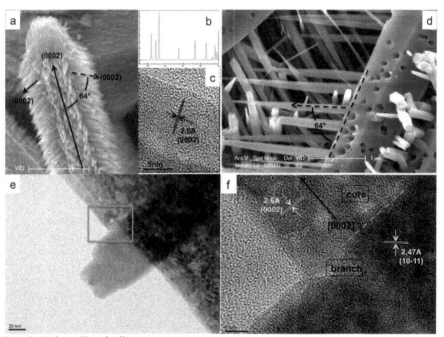

branches) about 40nm in diameter.

Figure 2. Analysis of branched structures: SEM micrograph (a) highlighting the secondary branches growing from the prismatic faces of the core rod. XRD pattern (b) confirmed wurtzite ZnO. HRTEM of a branch (c) uncovering the growth direction of the branch (i.e., along (0002).

Closer inspection via SEM (d) demonstrates the interfaces between the branches and the main rods as well as pores within the main rod. HRTEM micrograph (e) demonstrates the tip of the branch. Lattice measurements in TEM (f) confirm the growth direction of rods and reveals the interfaces between the core rods and the branches.

A SEM micrograph taken perpendicular to the long axis of the ZnO branched structure (Figure 2a) indicates the secondary rods grow from (10-10) planes, and the angle between the branches and [0001] direction in the core rod is about 64°. HRTEM analyses of the secondary rods (Figure 2c) revealed that the d-spacings of the branches correspond to ZnO [0002] direction, and confirmed the growth direction of the branching structure. High resolution SEM analyses confirmed the 64° angle between branch and [0001] direction, and also demonstrated pores on the prismatic surfaces of the primary rod (Figure 2d). The presence of these pores provides hints as to the growth mechanism of the branches, as these pores reveal locations of interfaces between the branches and the main rods. It is possible that during the rod growth, particles that are formed in solution may impinge on the growth front, forming twin boundaries with the main rod and acting as branch points. Similar observations have been made in highly branched TiO_2 nanostructures[37]. The presence of pores could have formed due to a number of reasons. One possibility is the etching of highly defective regions where branches had formed, yielding pores with hexagonal shapes[39]. Another possibility is the incomplete formation of branches or branches which may have initially formed and subsequently fractured due to growth stresses from the primary rod growth perpendicular to the prismatic (i.e., (10-10)) planes. Close observation of Figure 2d reveals that the pore diameter increases away from the core rod. This could indicate that branches are form early and continue to grow in diameter, but have limited growth near the core rod, as they are encased due to lateral growth of the primary rod. Further analyses of the pore formation are needed to explain this phenomenon.

TEM analysis (Figure 2f) of the branched ZnO nanowires confirmed the branch diameters (i.e., 20-40 nm). HRTEM (Figure 2e) highlights the tip of the branch in Figure 2f, on which there are ZnO nanoparticles observed attached on the branching rod. This implies the formation of the branches might be caused by attachment of ZnO nanoparticles that form twins[37, 40]. Additional HRTEM (Figure 2f) analysis was conducted at an interface (yellow rectangle area) of a branch and the core region. It demonstrates the lattice spacing of the core rod and the branched regions are 0.26nm and 0.247nm, respectively. HRTEM analyses demonstrated the core rod was growing along [0002] direction, and the branches that grow from the (10-10) plane also grow in the [0002] direction. Based on the crystallography of the ZnO system, the angle between [0002] and [10-11] is approximately 64°, which corresponds to our observations in the SEM and TEM micrographs (Figures 2a, d, and g). Thus, it is likely that nanoparticles attaching to growing core rods formed twins on the [10-11] plane[37, 38, 41] that initiated branching on these nanorods.

Growth Mechanism

We performed a time study (from 10 minutes to 20 hours) to determine the formation mechanism of the branched structures. Based on our observations (SEM), it is clear that at very short times (i.e., 10 min, Figure 3a), nanowires form concurrently with large numbers of nanoparticles. The sizes of the nanowires ranges from 50nm to 200nm, while the sizes of the nanoparticles are approximately 20nm. At longer durations (i.e., 30 min and 1 hr, Figures 3b and c, respectively), a primary hexagonal rod (d = 200nm for 30min, d = 400nm for 1hr) initially forms with minimal branching (< 3% of surface). By 5 and 10 hours (Figure 3e and f, respectively), a significantly higher density of branches has formed (20% of surface coverage) on the six prismatic (10-10) faces. Some of these branches are plate-like, since the needle-like branches (30nm diameter) can fuse together[28]. After 15hrs (Figure 3g and h), all six sides are filled with branches (95% surface coverage). XRD (Figure 3i) confirms that all the samples at different growth times are in the form of crystalline wurtzite-type ZnO (JCPDS 36-1415). The crystal size of the samples increased from 30nm (at 10 min) to approximately 60nm (at 20 min), and further increased with longer growth times (from 30min to 5hr). At significantly longer times (i.e., 10 – 20 hrs), the crystal size decreased (as observed by the increase in the full width at half maximum (FWHM) of the (100) ZnO peak). At short reaction durations (10 minutes), the sample is primarily composed of low aspect ratio ZnO nanoparticles, and thus the relative intensity of the (0002) peak is higher than samples formed at longer reaction times. With increasing reaction duration, the aspect ratio of these particles increases (i.e., nanorods form) as well as formation of branches and therefore, the relative intensity of (0002) peak decreases with reaction time. The TEM micrograph and electron diffraction pattern (Figure 3j) shows that the sample formed at 10 minutes consists of both thin (10 - 50 nm) ZnO nanowires and small (8 – 20 nm) ZnO nanoparticles. The presence of a large number of nanoparticles with nanorods suggests that branching events were triggered from twins formed from particles impinging on nanorod growth fronts[37]. The TEM bright field micrographs and electron diffraction pattern (Figure 3k and insets) analysis of the ZnO formed at 30 minutes indicates that thicker ZnO (~160 nm) rods are composed of smaller nanowires (10 nm) at the tip. HRTEM highlights the formation of the thick rod is due to oriented alignment of thin rods. At 5 hours reaction duration (Figure 3l), a composite structure forms consisting of ZnO rods (400 nm) and bound particles (30 nm). Electron diffraction analysis confirms both single crystal and polycrystalline ZnO in this sample. It is likely that the formation of these small particles are due to secondary nucleation in the solution after hydrolysis of EDA-bound Zn species, but further investigation is necessary.

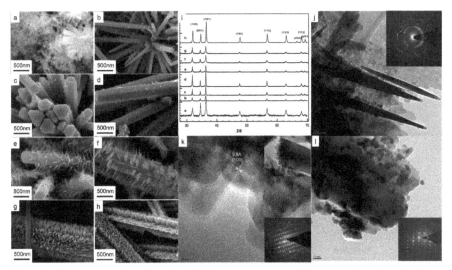

Figure 3. SEM micrographs of branched ZnO nanostructures at different reaction times: (a) 10 minutes, (b) 30 minutes, (c) 1 hour, (d) 2 hours, (e) 5 hours, (f) 10 hours, (g) 15 hours, and (h) 20 hours. XRD analyses (i) confirm wurtzite ZnO for all samples at various reactions times. TEM micrographs and electron diffraction reveals the nanostructures at (j) 10 minutes, (k) 30 minutes, and (l) 5 hours.

During the formation of the branched structures of the ZnO, it is clear that smooth hexagonal structures form first, with subsequent branching occurring with increasing time. The branches grow from the six prismatic (10-10) surfaces.

Regarding the nucleation and growth process, it is believed that EDA plays an essential role in the formation of ZnO branched structures[39, 42]. The interaction between EDA with either the Zn-species or the nucleating ZnO crystal can affect crystal nucleation and growth. To examine the role of EDA, the morphology of several different molar ratio of Zn to EDA has been compared. SEM data demonstrates (Figure 4) that with different ratios of Zn^{2+} to EDA, the morphologies of ZnO can be varied. Without EDA, we see the characteristic hexagonal rod-like structures of wurtzite ZnO, while increasing the ratio (Zn: EDA = 1:0.5), induces a very slight branching on the prismatic sides. The density of these branches is maximized at a ratio of Zn to EDA of 1:1. Increasing this ratio to 1:2 reduces the branching, while the highest ratio of 1: 20 yields no branching, and a very low aspect ratio (i.e., ZnO nanoparticles are produced instead of ZnO rods).

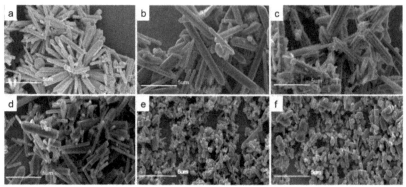

Figure 4. Effect of different molar ratios of Zn^{2+} to EDA on morphology. (a) $[Zn^{2+}]$: [EDA]=1:0, (b) $[Zn^{2+}]$: [EDA]=1:0.5, (c) $[Zn^{2+}]$: [EDA]=1:1, (d) $[Zn^{2+}]$: [EDA]=1:2, (e) $[Zn^{2+}]$: [EDA]=1:10, (f) $[Zn^{2+}]$: [EDA]=1:20. Samples were synthesized at 180°C, pH=14, [Zn]=50mM and [Zn]: [OH]=1:20.

The following reactions depict interactions of EDA with the zinc species in aqueous solutions[39, 43]:

$$Zn^{2+} + 4OH^- \rightarrow ZnO + 2OH^- + H_2O \qquad (1)$$

$$Zn^{2+} + 3NH_2(CH_2)_2NH_2 \leftrightarrow [Zn(NH_2(CH_2)_2NH_2)_3]^{2+} \qquad (2)$$

$$ZnO + NH_2(CH_2)_2NH_2 + H_2O \leftrightarrow [Zn(NH_2(CH_2)_2NH_2)_3]^{2+} + 2OH^- \qquad (3)$$

In the solution, a variety of zinc complexes are generated. Since there is a high concentration of OH- ions in the solution, the supersaturation of ZnO_2^{2-} is very high, leading to fast homogeneous precipitation and growth[19]. In the absence of EDA, all of the Zn^{2+} would be hydrolyzed (i.e., $[Zn(OH)_4]^{2-}$ and subsequently condensed to form ZnO nuclei. These nuclei subsequently grow into one-dimensional nanorods along the fastest growth direction [0001]. With the addition of EDA, there is a competition between the OH⁻ ions and EDA for the Zn^{2+} cation. Those cations reacted with OH- can participate in condensation reactions and form ZnO nuclei. Some of the remaining Zn^{2+} can be chelated by EDA, (a Zn-EDA complex forms when the molar ratio of EDA to Zn^{2+} is higher than 3[42]). This soluble Zn-EDA complex would not immediately participate in hydrolysis and condensation reactions. At temperatures greater than 95°C, the Zn-EDA complex will start to decompose[44] and the equilibrium condition of Equation (2) would shift to the left, resulting in an increase in the concentrations of Zn^{2+} and EDA. The free EDA can etch the top and side surfaces of already formed ZnO nanorods and create nucleation sites for secondary nucleation (equation 3)[39], while the released Zn^{2+} ions are susceptible to nucleophilic attack by OH- ions in the solution and serve as the precursor for secondary nucleation, which can be observed from Figure 2a. This would lead to the branches growing from the prismatic (10-10) surfaces and the (0002) surface along the fast growth [0002] direction. Thus at low concentrations of EDA (Zn^{2+}: EDA=1:0.5), there are only very few branches on the rod surfaces, while there are a high density of branches for Zn^{2+}: EDA=1:1. The higher concentration of EDA (Zn^{2+}: EDA=1:1) induces secondary nucleation (and potentially more nucleation sites) yielding a high density of branches. With even higher EDA concentrations, such as Zn^{2+}: EDA=1:10 or 1:20, EDA serves as reservoir for zinc cations[25]. Under certain conditions, EDA is able to compete with the hydroxide ion for the zinc cations[41]. Therefore, most of the zinc cations are inhibited from hydrolysis at low temperatures and high concentrations of EDA and will thus only be susceptible to hydrolysis at high temperatures. At high temperatures, the supersaturation level of Zinc cation is significantly higher resulting in a large number of smaller nuclei, which aggregate to form ZnO nanoparticles rather than long rods or branched structures.

Since the branched structures are affected by the presence of EDA in solution, it is unclear if EDA is incorporated into the crystal structure. FTIR was conducted on both smooth ZnO rods synthesized without EDA and branched ZnO structure synthesized with EDA, and the peaks are marked as shown in Figure 5. The strong peak at 470 cm^{-1} corresponds to E_2 mode of hexagonal ZnO, and the peak at 505 cm^{-1} corresponds to the oxygen deficiency and/or oxygen vacancy (VO) defect complex in ZnO[45]. The presence of a carboxylate group likely comes from carbon dioxide in the environment and the O-H stretch results from the presence of surface adsorbed water as well as Zn-OH surface species[46]. The absorbance bands at 1380 cm^{-1} and 3300 cm^{-1} were observed in the branched structured ZnO sample, revealing the peaks for C-N and N-H respectively, which would be indicative of the presence of EDA within, or on the surface of the branched structures. EDA is known to bind to the surfaces of ZnO [42, 47]. However, it is not clear that if the binding of EDA induces the branching event or if it is just present as a residual impurity in the solution.

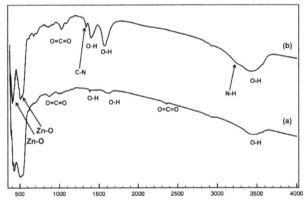

Figure 5. FTIR of smooth ZnO rods synthesized (a) without EDA and (b) branched ZnO structure synthesized with EDA.

CONCLUSIONS

We have developed a one-pot solution-based method for branched ZnO rods under relatively mild conditions. The branched structures were modified by using ethylenediamine as a mineralizer with different zinc precursor concentrations, and ratios of zinc precursor to EDA. In the process of producing the ZnO nanostructures, we found that EDA plays an inhibitory role, delaying complete hydrolysis and inducing secondary nucleation after the primary rods were growing, which yield branched structures. These branched structures likely form from nanoparticles, which impinge on the growing primary rod, forming the branches via [10-11] twinning. By understanding of the growth mechanism, we can design nanostructures with high surface area for potential enhanced performance in dye-sensitized solar cells.

REFERENCE

[1] Chen, X.; Li, C.; Graetzel, M.; Kostecki, R.; Mao, S. S., Nanomaterials for renewable energy production and storage. *Chemical Society Reviews.* **41**, 7909-7937 (1999).

[2] Gratzel, M., Conversion of sunlight to electric power by nanocrystalline dye-sensitized solar cells. *Journal of Photochemistry and Photobiology A: Chemistry.* **164**, *164*, 3-14 (2004).

[3] Gratzel, M., Photoelectrochemical cells. *Nature.* **414**, 338-344 (2001).

[4] O'Regan, B; Gratzeli M., A low-cost, high-efficiency solar cell based on dye-sensitized. *Nature.* **353**, 737-740 (1991).

[5] Green, M.A.; Emery, K.; Hishikawa, Y.; Warta, W.; Dunlop, E. D., Solar cell efficiency tables (version 39). *Progress in Photovoltaics: Research and Applications.* **20**, 12-20 (2012).

[6] Han, L.; Islam, A.; Chen, H.; Malapaka, C.; Chiranjeevi, B.; Zhang, S.; Yang X.; Yanagida, M., High-efficiency dye-sensitized solar cell with a novel co-adsorbent. *Energy & Environmental Science.* **5**, 6057-6060 (2012).

[7] Look, D.; Reynolds, D. C.; Sizelove, J. R.; Jones, R. L.; Litton, C. W.; Cantwell, G.; Harsch, W. C., Electrical properties of bulk ZnO. *Solid State Communications.* **105**, 399-401 (1998).

[8] Ozgur, U.; Alivov, Y. I.; Liu, C.; Teke, A.; Reshchikov, M. A.; Dogan, S.; Avrutin, V.; Cho, S. J.; Morkoc, H., A comprehensive review of ZnO materials and devices. *Journal of Applied Physics.* **98**, 041301-041301 (2005).

[9] Zhang, Q.F.; Dandeneau, C. S.; Zhou, X.; Cao, G., ZnO Nanostructures for Dye-Sensitized Solar Cells. *Advanced Materials.* **21**, 4087-4108 (2009).

[10] Anta, J. A.; Guillén, E.; Tena-Zaera, R., ZnO-Based Dye-Sensitized Solar Cells. *The Journal of Physical Chemistry C.* **116**, 11413-11425 (2012).

[11] Jagadish, C.; Pearton, S. J., *Zinc oxide bulk, thin films and nanostructures: processing, properties, and applications.* Hong Kong: Elsevier Science, 1-20 (2006).

[12] Ozgur, U.; Hofstetter, D.; Morkoc, H., ZnO Devices and Applications: A Review of Current Status and Future Prospects. *Proceedings of the Ieee.* **98**, 1255-1268 (2010).

[13] Elkhidir Suliman, A.; Tang, Y.; Xu, L., Preparation of ZnO nanoparticles and nanosheets and their application to dye-sensitized solar cells. *Solar Energy Materials and Solar Cells.* **91**, 1658-1662 (2007).

[14] Baxter, J.B.; Aydil, E. S., Nanowire-based dye-sensitized solar cells. *Applied Physics Letters.* **86**, 053114-053114-3 (2005).

[15] Leschkies, K.S.; Divakar, R.; Basu, J.; Enache-Pommer, E.; Boercker, J. E.; Carter, C. B.; Kortshagen, U.R.; Norris, D.J.; Aydil, E. S., Photosensitization of ZnO nanowires with CdSe quantum dots for photovoltaic devices. *Nano Letters*. **7**, 1793-1798 (2007).

[16] Han, J.; Fan, F.; Xu, C.; Lin, S.; Wei, M.; Duan, X.; Wang, Z. L., ZnO nanotube-based dye-sensitized solar cell and its application in self-powered devices. *Nanotechnology*, **21**, 405203-405209 (2010).

[17] Kisailus, D.; Schwenzer, B.; Gomm, J.; Weaver, J. C.; Morse, D. E., Kinetically controlled catalytic formation of zinc oxide thin films at low temperature. *Journal of the American Chemical Society*. **128**, 10276-10280 (2006).

[18] Xu, F.; Sun, L., Solution-derived ZnO nanostructures for photoanodes of dye-sensitized solar cells. *Energy & Environmental Science*. **4**, 818-841 (2011).

[19] Xia, Y.; Yang, P.; Sun, Y.; Wu, Y.; Mayers, B.; Gates, B.; Yin, Y.; Kim, F.; Yan, H., One-Dimensional Nanostructures: Synthesis, Characterization, and Applications. *Advanced Materials*. **15**, 353-389 (2003).

[20] Law, M.; Greene, L. E.; Johnson, J. C.; Saykally, R.; Yang, P., Nanowire dye-sensitized solar cells. *Nature materials*. **4**, 455-459 (2005).

[21] Jiang, C.Y.; Sun, X. W.; Lo, G. Q.; Kwong, D. L.; Wang, J. X., Improved dye-sensitized solar cells with a ZnO-nanoflower photoanode. *Applied Physics Letters*. **90**, 263501-263501 (2007).

[22] Kong, Y.; Yu, D.; Zhang, B.; Fang, W.; Feng, S., Ultraviolet-emitting ZnO nanowires synthesized by a physical vapor deposition approach. *Applied Physics Letters*. **78**, 407-409 (2001).

[23] Wu, J.; Liu, S., Low-temperature growth of well-aligned ZnO nanorods by chemical vapor deposition. *Advanced Materials*. **14**, 215-215 (2002).

[24] Huang, M.H.; Wu, Y.; Feick, H.; Tran, N.; Weber, E.; Yang, P., Catalytic Growth of Zinc Oxide Nanowires by Vapor Transport. *ChemInform*. **32**, 113-116 (2001).

[25] Govender, K.; Boyle, D. S.; Kenway, P. B.; O'Brien, P., Understanding the factors that govern the deposition and morphology of thin films of ZnO from aqueous solution. *Journal of Materials Chemistry*. **14**, 2575-2591 (2004).

[26] Ku, C.H.; Wu, J. J., Chemical bath deposition of ZnO nanowire,nanoparticle composite electrodes for use in dye-sensitized solar cells. *Nanotechnology*. **18**, 505706-505714 (2007).

[27] Ohyama, M.; Kouzuka, H.; Yoko, T., Sol-gel preparation of ZnO films with extremely preferred orientation along (002) plane from zinc acetate solution. *Thin Solid Films*. **306**, 78-85 (1997).

[28] Liu, B.; Zeng, H. C., Hydrothermal synthesis of ZnO nanorods in the diameter regime of 50 nm. *Journal of the American Chemical Society*. **125**, 4430-4431 (2003).

[29] De Yoreo, J. J.; Vekilov, P. G., Principles of crystal nucleation and growth. *Biomineralization*. **54**, 57-93 (2003).

[30] Lowenstam, H. A.; Weiner, S., *On biomineralization*. New York: Oxford University Press. 2-30 (1989).

[31] Yu, S. H.; Cölfen, H.; Hartmann, J.; Antonietti, M., Biomimetic Crystallization of Calcium Carbonate Spherules with Controlled Surface Structures and Sizes by Double-Hydrophilic Block Copolymers. *Advanced Functional Materials*. **12**, 541-545 (2002).

[32] Aizenberg, J.; Black, A. J.,; Whitesides, G. M., Oriented growth of calcite controlled by self-assembled monolayers of functionalized alkanethiols supported on gold and silver. *Journal of the American Chemical Society.* **121**, 4500-4509 (1999).

[33] Belcher, A. M.; Wu, X. H.; Christensen, R. J.; Hansma, P. K.; Stucky, G. D.; Morse, D. E., Control of crystal phase switching and orientation by soluble mollusc-shell proteins. *Nature.* **381**, 56-58 (1996).

[34] Gupta, A. K.; Gupta, M., Synthesis and surface engineering of iron oxide nanoparticles for biomedical applications. *Biomaterials.* **26**, 3995-4021 (2005).

[35] Cha, J. N.; Stucky, G. D.; Morse, D. E.; Deming, T. J., Biomimetic synthesis of ordered silica structures mediated by block copolypeptides. *Nature.* **403**, 289-292 (2000).

[36] Cao, G., *Nanostructures and Nanomaterials.* London: Imperial College Pr. 1-50 (2004).

[37] Li, D.; Soberanis, F.; Fu, J.; Hou, W.; Wu, J.; Kisailus, D., Growth mechanism of highly branched titanium dioxide nanowires via oriented attachment. *Crystal Growth & Design.* **13**, 442-428 (2013).

[38] Wang, B.; Shi, E.; Zhong, W., Twinning morphologies and mechanisms of ZnO crystallites under hydrothermal conditions. *Crystal Research and Technology.* **33**, 937-941 (1998).

[39] Sounart, T. L.; Liu, J.; Voigt, J. A.; Huo, M.; Spoerke, E. D.; McKenzie, B., Secondary nucleation and growth of ZnO. *Journal of the American Chemical Society.* **129**, 15786-15793 (2007).

[40] Zhang, Q.; Liu, S. J.; Yu, S. H., Recent advances in oriented attachment growth and synthesis of functional materials: concept, evidence, mechanism, and future. *Journal of Materials Chemistry.* **19**, 191-207 (2009).

[41] Zhao, F.; Zheng, J. G.; Yang, X.; Li, X.; Wang, J.; Zhao, F.; Wong, K. S.; Liang, C.; Wu, M., Complex ZnO nanotree arrays with tunable top, stem and branch structures. *Nanoscale.* **2**, 1674-1683 (2010).

[42] Gao, X.; Li, X.; Yu, W., Flowerlike ZnO nanostructures via hexamethylenetetramine-assisted thermolysis of zinc-ethylenediamine complex. *The Journal of Physical Chemistry B.* **109**, 1155-1161 (2005).

[43] Liu, Y.; Kang, Z. H.; Chen, Z. H.; Shafiq, I.; Zapien, J. A.; Bello, I.; Zhang, W. J.; Lee, S. T., Synthesis, characterization, and photocatalytic application of different ZnO nanostructures in array configurations. *Crystal Growth and Design.* **9**, 3222-3227 (2009).

[44] Lu, F.; Cai, W.; Zhang, Y., ZnO hierarchical micro/nanoarchitectures: solvothermal synthesis and structurally enhanced photocatalytic performance. *Advanced Functional Materials.* **18**, 1047-1056 (2008).

[45] Xiong, G.; Pal, U.; Serrano, J. G.; Ucer, K. B.; Williams, R. T., Photoluminesence and FTIR study of ZnO nanoparticles: the impurity and defect perspective. *physica status solidi (c).* **3**, 3577-3581 (2006).

[46] Tang, L. G.; Hon, D. N. S., Chelation of chitosan derivatives with zinc ions. III. Association complexes of Zn^{2+} onto O,N-carboxymethyl chitosan. *Journal of Applied Polymer Science.* **79**, 1476-1485 (2001).

[47] Liu, B.; Zeng, H. C., Room temperature solution synthesis of monodispersed single-crystalline ZnO nanorods and derived hierarchical nanostructures. *Langmuir.* **20**, 4196-4204 (2004).

A COMPARISON ON THE STRUCTURAL AND MECHANICAL PROPERTIES OF
UNTREATED AND DEPROTEINIZED NACRE

Maria I. Lopez[1], Po-Yu Chen[2], Joanna McKittrick[1,3], Marc A. Meyers[1,3]

[1] Materials Science and Engineering Program, University of California, San Diego, La Jolla,
CA92093, USA
[2] Department of Materials Science and Engineering, National Tsing Hua University, Hsinchu
30013, ROC
[3] Department of Mechanical and Aerospace Engineering, University of California, San Diego, La
Jolla, CA92093, USA

ABSTRACT

The contribution of the individual constituents of red abalone (*Haliotis rufescens*) to the strength of the nacre structure is investigated. Nacre sections were deproteinized to establish the contribution of the organic components. Tensile testing, scratch, and nanoindentation tests are performed on the isolated mineral constituent (deproteinized nacre) and the untreated nacre of red abalone shell. Specimens are characterized by scanning electron and atomic force microscopies to verify the deformation mechanisms. Results obtained from the isolated mineral validate the importance of the organic constituent, as the mechanical properties decline greatly when the organic component is removed. Scratch tests reveal the anisotropy of the material and the effects of the thick layers of protein (mesolayers) on the deformation behavior. This approach confirms the importance of the integrated structure to the overall mechanical behavior of nacre.

INTRODUCTION

In science and technology there is always a need for refining and improvements. Nature can provide excellent solutions to many of these difficulties. Understanding the property and structure relationship of biological materials by a materials science and engineering approach provides novel means of designing and processing synthetic materials.

In many cases, biological materials are a composite of biominerals and organics, which independently are quite weak [1]. Calcium carbonate, the mineral constituent of the abalone shell is quite brittle. However, when combined with an organic, nature creates a composite (nacre) that has a hierarchical, ordered structure with greatly improved mechanical properties.

The abalone nacre has various levels of organization ranging from the macro-structure to the nano-level [2]. The first level is the molecular structure of the chitin fibers that are the structural component of the intertile organic layers and of the atomic crystalline structure of the calcium carbonate phase, aragonite. The second level consists of the interface between the mineral tiles, which is composed of organic layers ~20 nm thick. In addition these interlayers are porous and allow mineral formations between adjacent tiles, known as mineral bridges [3,4]. The mineral bridges have a diameter of ~20-50 nm and a height of the interlayer. The third level consists of aragonite hexagonal tiles, with lateral dimensions of 8-10 μm and thickness of ~0.5 μm. These aragonite tiles are comprised of nanosized islands that arise due to the embedment of biopolymer [5,6]. The fourth level is the mesolayers, thick layers of biopolymer that are formed due to seasonal fluctuations [3,7-9]. The mesolayers are approximately ~200 μm thick and appear separating tile assemblages of approximately 0.1-0.5 mm thick. The fifth level of hierarchy is the entire structural geometry, including the hard outer calcitic layer, making it a two-layer armor system optimized for strength and toughness [10]. In this study, the primary focus will be on the second and third levels.

The structure-property relationship in abalone nacre has been intensively studied because it has the highest strength and toughness of any shells [1-36]. The arrangement of the parallel mineral tiles with the organic interface diminishes crack propagation as the crack has to travel along the organic layers creating a tortuous path, and accordingly the toughness and the work of fracture are enhanced. In addition, the structure is anisotropic which results in an orientation dependence of the mechanical properties. Moreover, because the hierarchical structure, different toughening mechanisms function at different levels suggesting the importance of understanding the mechanical properties at each level.

The objective of this investigation is to attain a better understanding of the structure-property relationship of the isolated constituents (e.g. isolated mineral and isolated organic component) in abalone nacre. When compared to the integrated (untreated) structure it can aid in determining the contributions of the different hierarchical levels and components. These results are significant to understand the important characteristics of abalone nacre to aid in improving the latest attempts to produce novel nacre-inspired materials.

MATERIALS AND METHODS

Deproteinization

Removal of all organic material from the nacre was performed by submerging the specimen in a basic solution. For the scratch and nanoindentation specimens, deproteinization was done by immersing it in a 5.2wt% sodium hypochlorite solution (NaClO) at 20 °C with constant mixing for a period of 12 days (where the solution was replaced daily). Due to the delicate nature of the nacre pucks for tensile testing, a less aggressive solution, 0.5N sodium hydroxide (NaOH) at 20°C for 10 days under constant, gentle shaking (Figure 1b) was utilized for the deproteinization of the nacre pucks for the tensile experiments. Deproteinization resulted in a separation of the sample where a mesolayer was present. The distance between mesolayers varies greatly between specimens; two mesolayers can be from 0.1µm to 1mm apart. Thus, when the removal of the organic constituent occurred the nacre pucks would separate along the mesolayers, yielding in samples of different thicknesses (0.1µm to 1mm thick).

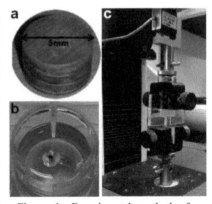

Figure 1: Experimental methods for tensile testing of deproteinized nacre pucks. a) Drilled puck specimen. b) Puck in deproteinizing solution. c) Mount setup for tensile testing.

Shell sectioning

Sectioning for tensile testing of deproteinized nacre was performed from two fresh abalone shells that were previously held and raised in an open water tank at the facility at the Scripps Institution of Oceanography, La Jolla, CA. The calcitic layer was removed via wet grinding, leaving only the nacreous layer. The samples were prepared by drilling cylindrical pucks of nacre, 5 mm in diameter, using a diamond coring drill (Figure 1a). Care was taken to make lateral surfaces perpendicular to the concavity of the surface of the shell to ensure that the inner nacre layers are as parallel to the ends of the cylindrical specimen as possible. Specimens where then ground and polished to create a flat surface. The thickness of these specimens varied from 0.3-3 mm.

For the scratch and nanoindentation specimens, nacre sections (3 cm x 3 cm x 0.3 cm) were cut using a diamond blade. Untreated specimens were directly mounted and polished. Deproteinized specimens were polished prior to and mounted after deproteinization. These specimens were prepared to be tested and characterized in two directions: top surface and in cross-section (Figure 2).

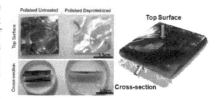

Figure 2: Nacre sectioning and mounting for scratch testing and nanoindentation.

Mount Setup

For the tensile testing of the nacre pucks, a setup was created to decrease damage. Once the organic constituent is removed, the nacre becomes brittle and fragile. To reduce any pre-loading prior to testing, the pucks were mounted in an acrylic setup that allowed gripping and handling of the sample (Figure 1c). In this setup, the tensile load was applied perpendicular to the tiles.

Mechanical Testing

Tensile testing was performed in a tabletop desktop Instron 3342 system at strain rates of $10^{-2}s^{-1}$. Nanoscratching was performed utilizing a CSM Nano Scratch Tester specially suited to characterize practical adhesion failure of thin films and coatings, with a typical thickness below 800 nm. Samples were tested by applying a progressive load up to 1000 mN for specimens tested in cross-section, and by applying a progressive load up to 600 mN for specimens tested on the top-surface. The scratch length varied from 2-3 mm depending on the available surface area. At least six high-quality scratches were performed on each specimen. The fracture surfaces of all the specimens were gold-platinum coated and observed in a FEI SFEG Ultrahigh resolution scanning electron microscope (SEM). Nanoindentation was performed using a Hysitron nanoindentation system in various regions of the untreated and deproteinized nacre at loads ranging from 300 mN to 500 mN. Indented specimen was observed sequentially by atomic force microscope (AFM).

RESULTS AND DISCUSSION

Imaging of Deproteinized Nacre

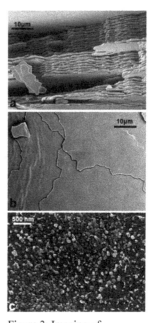

Figure 3: Imaging of deprotenized nacre pucks.
a) Cross-section showing the ~500 nm tiles.
b) 'Birds-eye-view' of fracture surface showing three different tile layers.
c) Nanoasperities covering the surface of the tiles.

Figure 3 shows the cross-sectional view of the nacre after deproteinization. It can be noted that the mineral tiles remain completely intact retaining their ~500 nm thickness and shape (Figure 3a). Subsequent to tensile testing, the fracture surface of the puck was observed via SEM. Figure 3b show one of these surfaces (taken as a 'birds-eye-view'). Different tile layers are peeled off as the load is applied. Closer inspection of this top surface (Figure 3c) reveals nanoasperities that cover the entire tile face with a uniform distribution.

Previous experimentation done on demineralized nacre [36], reveal the structure of the isolated organic material as a porous one composed by a network of fibers. Comparing the imaging of the isolated organic material and the isolated mineral allow for interesting conclusions (Figure 4). The pores found within the organic interlayer are hypothesized to enable the formation of mineral bridges between adjacent tile layers. Because the imaging is done directly on the fracture surface, we can presume that some of the nanoasperities are actually fractured mineral bridges. The nanoasperities and the holes within the organic matrix sheet were measured. On average, the radius of the pores found in the organic interlayer is ~20 nm. In comparison, the average radius of the nanoasperities was found to be ~33 nm. This difference in diameter size might be due to the relaxation of the membrane as the material is demineralized. As the mineral is removed the pores in the organic interlayers are no longer under stress and thus reduce in size. Past investigations have focused in defining the role of these mineral bridges and how they correlate to the structure of the organic interface [29-32]. Mineral bridges appear as circular columns with diameters 25-55 nm [31, 32], while the pores exhibit a diameters 5-50 nm [33,34]. Current results show an average diameter than falls on the larger end of the previously reported values on the nanoasperities, however, pore diameter measurements fit well with previous results [28].

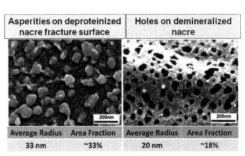

Asperities on deproteinized nacre fracture surface		Holes on demineralized nacre	
Average Radius	Area Fraction	Average Radius	Area Fraction
33 nm	~33%	20 nm	~18%

Figure 4: Comparison of the isolated organic material and the isolated mineral showing the differences in average diameter of asperities found on deproteinized nacre and the holes found on demineralized nacre.

However, it has been hypothesized that not all nanoasperities connect to form a mineral bridge. Previous studies suggest that in many cases the asperities only protrude [35]. The difference in the surface area covered by the nanoasperities and the area comprised by the holes in the membrane of the organic interlayer agrees with these previous results. Nanoasperities cover ~33% of the surface of the mineral tiles, compared to the area provided by the pores, which is estimated to be ~18%. Furthermore, Song and Bai [30] proposed that the average density distribution of mineral bridges vary – higher in the interior compared to the edges. In contrast, the current observations show a uniform distribution of nanoasperities on the surfaces of the tiles.

Tensile Tests of Deproteinized Nacre Pucks

Figure 5 shows the Weibull distribution of the deproteinized nacre pucks tested under tension with load perpendicular to layers. The 50% failure probability occurs at ~ 0.325 MPa, a low value, particularly when compared to the untreated nacre which shows a 50% failure probability at ~ 4.2 MPa.

Mineral bridges are believed to enhance stiffness, strength, and fracture toughness of the organic matrix by enhancing the crack extension pattern in nacre [29]. The theoretical strength of the mineral bridges is ~ 3.3 GPa [15]. If it is assumed that 18% of the surface area of each (approximating from the porous area in the organic intertile membrane) tile is covered by mineral bridges on samples 5 mm in diameter, the theoretical strength (3.3 GPa), far higher than what is measured. This can be due to several factors. There is likely an organic phase surrounding the mineral nanograins in the mineral bridges that deteriorated during the deproteinization process. Additionally, there is also the possibility that some of the mineral bridges were damaged or broken, previous to testing, lowering the strength values.

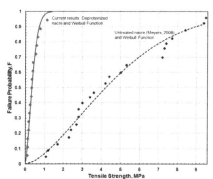

Figure 5: Weibull distribution of tensile strength perpendicular to layered structure of deproteinized nacre compared to results on whole nacre [15].

Nanoscratch Test

Figure 6 shows selected plots on the various sets of tests: a,b top surface untreated and deproteinized, respectively; b,c cross-section untreated and deproteinized, respectively. Interesting features can be noticed. When tested on the top surface, as expected, the deproteinized nacre fractures at lower loads than the untreated nacre; where major fractures began at the initial loading (3 mN) and fractured completely at loads lower than <100 mN. Force plots for this specimen do not show an explicit point of fracture and almost no resistance to scratching. In comparison the untreated nacre (tested on the top surface) exhibits a evident fracture limit, on average at ~27 mN. Furthermore, the anisotropic behavior can be noticed in the scratch tests. Compared to when tested on the top surface, when tested in the cross-section, the deproteinized nacre exhibited more of a resistance to scratching and demonstrated an explicit breaking point at ~120 mN.

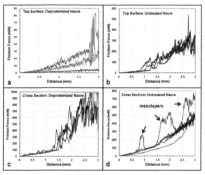

Figure 6: Microscratch force plots of: a) untreated nacre tested along the top surface, b) deproteinized nacre tested along the top surface, c) untreated nacre tested along the cross-section d) deproteinized nacre tested along the cross-section.

Additionally, when tested in cross-section, the untreated nacre does not show a precise frictional force limit; there is a gradual cracking which is more evident by SEM observations, discussed below. Furthermore, on untreated samples, mesolayers have an effect on the behavior and the frictional force. When tested in cross-section, mesolayers were encountered in various locations, when the indented tip meets a mesolayer, the scratch is deflected from its original path and follows through the mesolayer. It is also noticeable from the plot that the measured force increases as the mesolayer is encountered giving an increased the resistance to motion, suggesting mesolayers add plowing friction.

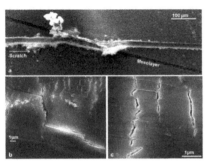

Figure 7: a) SEM micrograph of a scratch path encountering a mesolayer in untreated nacre. b) Fracture path in untreated nacre. c) Fracture path in deproteinized nacre.

When tested on the top surface, scratch profiles suggest that when the mineral is isolated there is very little resistance to fracture. This behavior corroborates similar behavior demonstrated by Bezares et al. [6], where nanoindentation was performed on heat-treated specimens where there was a loss in intra-tile protein. Heat-treated specimens appeared compacted, similar to heat-treated sand where grains begin to fuse together. However, in untreated specimens, the content of the organic component within the tiles that forms the 'nano-grain' structure the aragonite tiles in nacre, causing micro-crack deflection and crack blunting.

Moreover, there is an evident effect by all organic components in the material. The most evident feature of this is the mesolayers. Figure 7a shows an SEM micrograph of a scratch path encountering a mesolayer. As the scratch hits the mesolayer, the scratch path is deflected to follow the interface along the mesolayer. Furthermore, observations of the cross-section of tested untreated and deproteinized nacre show the impact of the organic interlayer. In the untreated nacre (Figure 7b) the crack propagates in a tortuous and step-wise; a much more complex path compared to that of deproteinized nacre. The crack is indeed deflected and arrested due to the successive combination of mineral and organic layers. The SEM image of the tested deproteinized nacre (Figure 7c) show that the crack growth precedes with a relatively unimpeded manner compared to the untreated nacre. Fractures occur through the tile and not necessarily following a predicted path. This difference in behavior again re-instates the importance of the organic component.

Nanoindentation experiments

Figure 8 shows AFM observations of an indentation on the center of the tile of untreated nacre. There is little, if any, crack propagation in untreated nacre when indented. For example in Figure 8a, the indent did not cause a crack, while in Figure 8b the indent created a crack. In Figure 8b, as the crack propagates, it reaches the edge it causing an aperture at the tile interface. Figure 9 shows the nanoindentation profile of deproteinized nacre. The surface of the deproteinized nacre is very different than that of the untreated nacre. The surface is rough and uneven. The tiles in nacre are known to contain embedment of organic material within the mineral [5]. With the deproteinization process, further than removing the organic interlayers,

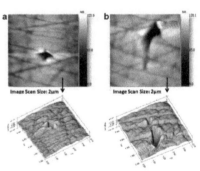

Figure 8: Nanoindentation profile of untreated nacre. a) Indentation contained within the tile. b) Indentation causes a crack to propagate which causes aperture at tile interface.

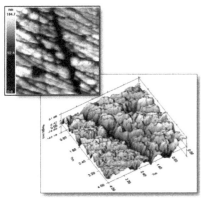

Figure 9: AFM observation of nanoindentation profile of deproteinized nacre. Extreme roughness and granular features conceal indent.

this embedded organic material within the mineral is also removed causing the gaps* in the surface of the tiles. The deproteinized nacre is extremely granular, thus when the diamond tip indents the surface, the indentation mark is concealed within the gaps. It is extremely difficult to observe any crack formation or propagation. These differences in nanoindentation profiles of the untreated nacre and deproteinized nacre agree with previous nanoindentation results by Bezares et al. [6]. Nanoindented untreated nacre showed pileups around edges, while heat-treated nacre was described as granular and loosely compacted. In this study, when the organic component is removed it also alters the structure of the mineral tiles, suggesting that in fact there is an embedment of organic constituent inside the tiles that impacts the mechanical properties.

CONCLUSIONS

The principal conclusions that can be drawn from the current research are:
1. From the mechanical testing of the deproteinized nacre we conclude that the behavior of the material, in particular the strength, is far below that of its theoretical strength. Even though the organic matrix accounts for only 10 vol% of nacre, when it is removed the strength is reduced by ~92% compared to whole nacre. This may be due to not only the removal of actual organic layers that have an effect of the weakening of the material, but also the removal of the organic material embedded within the mineral and/or bridges.
2. Some of the nanoasperities correspond to grains, and not mineral bridges. Distribution and density of the nanopores within the organic interlayer correspond to a better estimate of the number of mineral bridges.
3. When scratched on the top surface, deproteinized nacre fractures at a lower load and it does not show an explicit frictional force limit, compared to that of untreated nacre which exhibits an evident fracture limit (on average at ~27 mN). Furthermore, when scratched mesolayers (in untreated nacre) have an effect on the fracture behavior, adding a plowing force.
4. Scratch results also show the anisotropic behavior of nacre. In the scratch experiments nacre exhibits a higher resistance to failure when tested in cross-section (when the surface is along the tile layers) than from the top surface, in both, untreated and deproteinized specimens.
5. Nanoindentation results further reveal the effect of the loss of the organic constituent. Penetration in untreated nacre showed aperture at the tile interface while penetration in deproteinized nacre demonstrated the granular nature of the specimen due to the loss of the organic material.

ACKNOWLEDGEMENTS

This research is supported by National Science Foundation Grant DMR 1006931, NSF EAPSI Grant 1108531 and a Ford Foundation Pre-Doctoral Fellowship. Thanks go to Professor Duh, Yu-Chen, Hsien-Wei, Prof. JW Lee and his research group for help and access of equipment at National Tsing Hua University and National Taipei University of Technology in Taiwan.

REFERENCES

1. Meyers MA, Chen PY, Lin AYM, Seki Y. Biological materials: structure and mechanical properties. Prog Mater Sci 2008;53:1-206.
2. Meyers MA, Chen PY, Lopez MI, Seki Y, Lin AY. Biological materials: a materials science approach. J. Mech Behav Biomed Mater 2011;4:626-57.
3. Schaffer TE, IonescuZanetti C, Proksch R, Fritz M, Walters DA, Almqvist N, et al. Does abalone nacre form by heteroepitaxial nucleation or by growth through mineral bridges? Chem Mater 1997;9:1731-40.
4. Meyers MA, Lim CT, Li A, Nizam BRH, Tan EPS, Seki Y, et al. The role of organic intertile layer in abalone nacre. Mat Sci Eng C-Mater 2009;29:2398-410.
5. Rousseau M, Lopez E, Stempfle P, Brendle M, Franke L, Guette A, et al. Multiscale structure of sheet nacre. Biomaterials 2005;26:6254-62.
6. Bezares J, Peng Z, Asaro RJ, Zhu Q. Macromolecular structure and viscoelastic response of the organic framework of nacre in *Haliotis rufescens*: a perspective and overview. Journal of Theoretical Applied Mechanics 2011;38:75-106.
7. Menig R, Meyers MH, Meyers MA, Vecchio KS. Quasi-static and dynamic mechanical response of *Haliotis rufescens* (abalone) shells. Acta Mater 2000;48:2383-98.
8. Lin A, Meyers MA. Growth and structure in abalone shell. Mat Sci Eng a-Struct 2005;390:27-41.
9. Su XW, Belcher AM, Zaremba CM, Morse DE, Stucky GD, Heuer AH. Structural and microstructural characterization of the growth lines and prismatic microarchitecture in red abalone shell and the microstructures of abalone "flat pearls". Chem Mater 2002;14:3106-17.
10. Cranford SW, Buehler MJ. Biomateriomics. Springer Durdrecht Heidelberg New York London 2012.
11. Zaremba CM, Belcher AM, Fritz M, Li YL, Mann S, Hansma PK. Critical transitions in the biofabrication of abalone shells and flat pearls. Chem Mater 1996;8:679-90.
12. Lin AYM, Meyers MA, Vecchio KS. Mechanical properties and structure of *Strombus gigas, Tridacna gigas,* and *Haliotis rufescens* sea shells: a comparative study. Mat Sci Eng C-Bio S 2006;26:1380-9.
13. Meyers MA, Lin AYM, Seki Y, Chen PY, Kad BK, Bodde S. Structural biological composites: an overview. Jom-Us 2006;58:35-41.
14. Lin AYM, Chen PY, Meyers MA. The growth of nacre in the abalone shell. Acta Biomater 2008;4:131-8.
15. Meyers MA, Lin AYM, Chen PY, Muyco J. Mechanical strength of abalone nacre: role of the soft organic layer. J Mech Behav Biomed 2008;1:76-85.
16. Kobayashi I, Samata T. Bivalve shell structure and organic matrix. Mat Sci Eng C-Bio S 2006;26:692-8.
17. Lin AYM, Meyers MA. Interfacial shear strength in abalone nacre. J Mech Behav Biomed 2009;2:607-12.
18. Sarikaya M, Gunnison, K. E., Yasrebi, M. and Aksay, J. A. Mechanical property – microstructural relationships in abalone shell. In Materials Synthesis Utilizing Biological Processes. Materials Research Society 1990;174:109-66.

19. Nakahara H, Kakei, M., and Bevelander, G. Electron microscopic and amino acid studies on the outer and inner shell layers of *Haliotis rufescens*. Venus Jpn J Malac 1982;41:33-46.
20. Sarikaya M, Aksay IA. Nacre of abalone shell: a natural multifunctional nanolaminated ceramicpolymer composite material. Results Probl Cell Differ 1992;19:1-26.
21. Nukala P.K.V.V., Simunovic S. A continuous damage random thresholds model for simulating the fracture behavior of nacre. Biomaterials 2005;26:6087-98.
22. Li XD, Xu ZH, Wang RZ. In situ observation of nanograin rotation and deformation in nacre. Nano Lett 2006;6:2301-4.
23. Currey JD, Kohn AJ. Fracture in crossed-lamellar structure of conus shells. J Mater Sci 1976;11:1615-23.
24. Currey JD. Mechanical-properties of mother of pearl in tension. P Roy Soc B-Biol Sci 1977;196:443.
25. Laraia VJ, Heuer AH. Novel composite microstructure and mechanical-behavior of mollusk shell. J. Am Ceram Soc 1989;72:2177-9.
26. Barthelat F, Espinosa HD. An experimental investigation of deformation and fracture of nacre mother of pearl. Exp Mech 2007;47:311-24.
27. Barthelat F, Tang H, Zavattieri PD, Li CM, Espinosa HD. On the mechanics of mother-of-pearl: A key feature in the material hierarchical structure. J Mech Phys Solids 2007;55:306-37.
28. Tang H, Barthelat F, Espinosa HD. An elasto-viscoplastic interface model for investigating the constitutive behavior of nacre. J Mech Phys Solids 2007;55:1410-38.
29. Song F, Bai YL. Mineral bridges of nacre and its effects. Acta Mech Sinica 2001;17:251-7.
30. Song F, Zhang XH, Bai YL. Microstructure and characteristics in the organic matrix layers of nacre. J. Mater Res 2002;17:1567-70.
31. Song, F., Soh, A. K. and Bai, Y. L. Structural and mechanical properties of the organic matrix layers of nacre. Biomaterials 2004; 24, 3623-3631
32. Gries K, Kroger R, Kubel C, Fritz M, Rosenauer A. Investigations of voids in the aragonite platelets of nacre. Acta Biomater 2009;5:3038-44.
33. Bezares J, Asaro RJ, Hawley M. Macromolecular structure of the organic framework of nacre in *Haliotis rufescens*: Implications for growth and mechanical behavior. J Struct Biol 2008;163:61-75.
34. Bezares J, Asaro RJ, Hawley M. Macromolecular structure of the organic framework of nacre in *Haliotis rufescens*: Implications for mechanical response. J Struct Biol 2010;170:484-500.
35. Checa AG, Cartwright JH, Willinger MG. Mineral bridges in nacre. J Struct Biol 2011;176:330-9.
36. Lopez MI, Meza-Martinez PE, Meyers MA, Organic Interlamellar layers, Mesolayers, and Mineral Nanobridges: Contribution to Strength in Abalone (Haliotis rufecens) Nacre. Acta Biomaterialia (Submitted, 2013)

REINFORCING STRUCTURES IN AVIAN WING BONES

E. Novitskaya[1*], M.S. Ribero Vairo[1,2], J. Kiang[1], M.A. Meyers[1], J. McKittrick[1]

[1]University of California, San Diego, Department of Mechanical and Aerospace Engineering and Materials Science and Engineering Program, 9500 Gilman Dr., La Jolla, CA 92093, USA
[2]Universidad Nacional de Cuyo, Centro Universitario, Facultad de Ingenieria and ITIC, M5502JMA, Mendoza, Argentina

ABSTRACT
    Nearly all species of modern birds are capable of flight; therefore mechanical competency of appendages and the rigidity of their skeletal system should be optimized. Birds have developed extremely lightweight skeletal systems that help aid in the generation of lift and thrust forces as well as helping them maintain flight over, in many cases, extended periods of time. The humerus and ulna of different species of birds (flapping, flapping/soaring, flapping/gliding, and non-flying) have been analyzed by optical microscopy and mechanical testing. The reinforcing structures found within bones vary from species to species, depending on how a particular species utilizes its wings. Interestingly, reinforcing ridges and struts have been found within certain sections of the bones of flapping/soaring and flapping/gliding birds (vulture and sea gull), while the bones from the flapping bird (raven) and non-flying bird (domestic duck) did not have supporting structures of any kind. The presence of these reinforcing structures increases the resistance to torsion and flexure with a minimum weight penalty, and is therefore of importance in flapping/gliding birds. Vickers hardness testing was performed on the compact section of the bones of all bird species. The data from the mechanical testing were compared with microstructural observations to determine the relevance behind the reinforcing structures and its mechanical and biological role. Finite element analysis was used to model the mechanical response of vulture ulna in torsion.

1. INTRODUCTION
    Mechanical engineering is an interdisciplinary field that encompasses studies such as solid mechanics and material science. By understanding and using the core concepts behind these studies, mechanical engineers are able to analyze, design, manufacture, and maintain mechanical systems. Biomimetics is the application of the structure and function of biological systems for the design of new machines and materials, and is emerging as a new area of interest that opens up a completely different view on mechanical engineering.
    In nature, excellent examples of engineering solutions are found. These engineering solutions have been perfected over millions of years of evolution. By studying and understanding lessons from nature, new or better designs of materials and structures can be made[1]. For biomimetics, it is important to have a clear understanding of biology.
    Outstanding examples of structural adaptation are avian wing bones. This has been recognized close to one hundred years ago by Darcy Thompson[2]. These bones have evolved over time to allow the birds to achieve and maintain flight. One adaptation is the fusion of several bones into a single ossification. The carpometacarpus (blade-like structure of wrist and hand bones) is an example of fused bird bones, which helps to provide additional strength to the wing[3,4]. By fusing the bones, the total number of bones found within a bird skeletal system is far less than that of other terrestrial vertebrates[3,4]. Additionally, the skeleton becomes much more lightweight as well as rigid. Another adaptation for flight is that many of the bones are hollow or semi-hollow. The hollow bones help to

* Corresponding author, phone: 858-534-5513, fax: 858-534-5698, e-mail: eevdokim@ucsd.edu

offset the high-energy cost of flight[4]. In addition, air pockets (pneumatic foramens[3]) often form within the hollow or semi-hollow bones of birds (e.g., humerus and skull). These pockets are part of the "flow-through ventilation" system that avian species use to move air through their lungs, forming pneumatic bones[5,6].

The bird wing consists of several main bones such as humerus ('upper arm'), radius and ulna ('forearm'), carpometacarpus to form the 'wrist' and 'hand' of the bird, and the digits ('fingers') that are fused together. The main flight muscles of the breast are attached only to a humerus bone; therefore this bone has an important role of bearing the large forces during the flight[4]. The ulna is one of two bones that support the midsection of the wing. For the flying birds the humerus is usually shorter and thicker compare to ulna, since it needs to withstand larger forces during the flight[4]. In addition, the bones of various avian species have microstructural features (osteonal structure, Haversian canals, and lacunae) similar to other mammalian long bones[6].

Some birds achieve and maintain flight by flapping their wings as well as soaring through the air (flapping/soaring birds[8], e.g. vultures, eagles); others are flapping and gliding (flapping/gliding birds[8], e.g. sea gulls, pelicans)). Furthermore, some birds are able to alternate between flapping their wings with only periodic gliding (flapping birds, e.g. ravens, crows). Some birds are flightless due to environmental and habitat conditions of their growth (e.g. domestic ducks, emus).

It has been shown that reinforcing structures are found within wing bones in the places of maximum torsional and bending moments[4,9]. These structures (struts) mostly appear at the places "in need", preventing the buckling of bone walls due to internal loads[9]. Another type of supporting structure (ridges) was found inside the wing bones of flapping/soaring and flapping/gliding birds[9]. These structures are similar to ship supporting trusses which have a function of optimization and redistributing of external stresses. A detailed analysis of reinforcing structures (both struts and ridges), and mechanical properties of two wing bones, the humerus and ulna, from flapping, flapping/soaring, flapping/gliding, as well as a non-flying bird was performed in this study. Additionally, a first approach on the understanding of the mechanical behavior of bird wing bones in torsion using finite element analysis (FEA) is presented in this work.

2. MATERIALS AND METHODS

2.1 Sample preparation

Bone samples from ulna and humerus were gathered from a flapping/soaring bird (the Turkey Vulture, *Cathartes aura*), a flapping/gliding bird (the California Gull, *Larus californicus*), a flapping bird (the Common Raven, *Corvus corax*) and a non-flying bird (the Pekin Duck, *Anas peking*). Bones were stored in ambient dry condition at room temperature and normal humidity.

2.2 Mineral content

The mineral content of bird bones was measured by weight. First, cleaned samples (about 1 cm height cylinders) were submerged in Hank's balance saline solution for 24 hr for rehydration. Then, the water content was evaporated by heating the bones in an oven at 105°C for four hours. The weights of the individual samples were measured before and after the heating processes, providing the information about water content of the bones. Next, bone samples were further heated in an oven for 24 hours at 550°C to eliminate the proteins. The weights of the individual samples were measured before and after the heating process. Weight percent of minerals (wt.%) was calculated by dividing the weight after by the weight before heating.

2.3 Structural characterization

Cross-sections of ulnae and humeri were prepared for each bird species; next they were embedded into epoxy and polished for future optical observation and hardness testing. Samples from all four species were analyzed by optical microscopy using Zeiss Axio imager equipped with CCD camera (Zeiss Microimaging Inc., Thornwood, NY), and Keyence VHX1000 microscope (KEYENCE America, Elmwood Park, NJ).

2.4 Image processing

An image processor, ImageJ, was used to analyze the porosity of the bone samples, similar to the porosity analysis by Manilay et al.[10] The Haversian system (including vascular channels and Volkmann's canals), and lacuna spaces were the pore types used for the porosity calculations. Porosity values were calculated dividing the sum of the areas of the pores by the total area of the image.

2.5 Micro-computed tomography (μCT)

A section from a distal part of Turkey Vulture ulna was scanned on a micro-computed tomography (μCT) scanner, Skyscan 1076 (Kontich, Belgium). Bone was scanned inside a dry plastic tube. Imaging was performed at 36 μm isotropic voxel sizes applying an electric potential of 70 kV and a current of 200 μA, using a 0.5 mm aluminum filter. Images and 3-dimensional (3D) rendered models were developed using Skyscan's DataViewer and CTVox software.

2.6 Hardness testing

Hardness from all four species was measured using a LECO M-400-H1 hardness testing machine equipped with a Vickers hardness indenter. The cross-sectional bone samples embedded into epoxy were analyzed at different locations to determine the overall hardness distribution across a single cross-section of the bone. Hardness values were averaged from 20 micro-indentations. A load of 10 g_f was used to indent the exposed surfaces.

2.7 Statistical analysis

One-way ANOVA analysis was performed to determine significant differences between the hardness data for humerus and ulna of the same bird species, and among species. The criterion for statistical significance was $p < 0.05$.

2.8 Finite element analysis (FEA)

A small section of the vulture ulna (a cylinder with 13 mm in height, 9 mm in diameter, and 0.8 mm in wall thickness) from the distal end (Figures 1, 2) was analyzed in torsion by finite element program LS-DYNA (Version 971)[11]. The geometry was obtained from μ-CT images, and discretized with 151.083 tetrahedral solid elements with aspect ratios between 1.03 and 5.96. A linear elastic isotropic model with a homogeneous distribution of properties was considered. The elastic properties were taken from the literature[9] and from mechanical testing assuming isotropy as: Young's modulus E = 20 GPa, Poisson's ratio (v) = 0.3. The geometry discretization and boundary conditions are shown in Figure 1. Nodes in the bottom of the samples were constrained to a zero displacement, while the top ones were subjected to an angular velocity of 0.04 rad s^{-1}, with a final rotation of 0.2 rad to avoid geometrical nonlinearities during the simulation. The rotational axis was coincident with the torque vector **T** (Figure 1b), and applied along the center of inertia of the whole sample. No inertia effects were considered in the simulation. The final result was shown as a von Mises stress distribution.

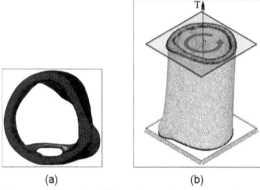

(a)   (b)

Figure 1. (a) The μCT image of the section of the Turkey Vulture ulna (top view). (b) Geometry discretization and boundary conditions for the section. The planes on the top and bottom of the sample are defined by the end nodes.

3. RESULTS AND DISCUSSION
The left wing of the Turkey Vulture is shown in Figure 2 as an example of the overall bone configuration of a bird wing.

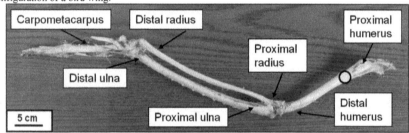

Figure 2. Left wing of the Turkey Vulture. Proximal and distal ends of humerus, ulna, and radius are marked. The yellow dot shows the point of maximum bending and torsional moments carried by the humerus in flight[9].

Proximal (closest to body) and distal (farthest from body) ends of humerus, ulna, and radius are marked. The yellow dot shows the point of maximum bending and torsional moments carried by the humerus in flight[9]. A basic comparison of the sizes of ulnae and humeri for flying (vulture) and non-flying (duck) birds are shown in Figure 3. The proximal and distal ends of the bones are shown. It is clear that for flying birds ulna is longer compare to humerus, while it is opposite for non-flying birds[4,9]. The ratio between the lengths of humeri and ulnae was similar for the flying birds (0.8-0.9), while it was ~1.5 for the non-flying bird. The ways birds use their wings, as well as torsional and flexure stresses that bones experience during the flight for the former case and lack of these events for the latter one are the main reasons for this fact.

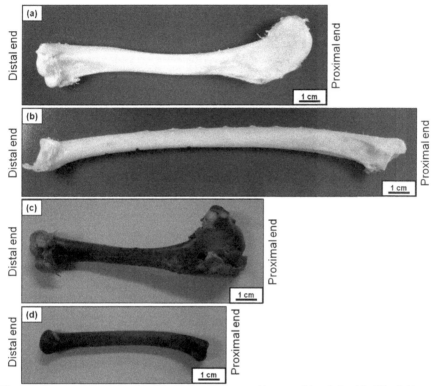

Figure 3. Humerus (a) and ulna (b) of the Turkey Vulture, and humerus (c) and ulna (d) of the Pekin Duck.

Microstructural analysis was performed on the entire cross-sections of ulnae and humeri for all species; representative images of ulnae cross-sections are shown in Figure 4. It is clear that thickness of the bone wall is not uniform for all flying birds due to presence of external pressure and stress distribution on them during the flight. Additionally, bones from flapping/soaring and flapping/gliding birds have ovalized cross-sections, while bones from flapping and non-flying birds have more circular cross-sections. The ulna from non-flying bird (duck) has the most circular cross-section. Furthermore, ulnae of the vulture and the gull have the reinforcing structures (struts), while ulnae of the raven and the duck lack them (Figure 4). These struts are thought to be in the places "in need" to support the bone against extensive ovalization during torsional and flexure loading[9]. The ovalization usually appears in bones that are subjected to high bending moments during flight. The ovalization changes the cross-sectional shape and weakens the whole bone structure resulting in unstable elastic deformation[9]. This weakening can be corrected by the presence of relatively long struts that oriented at 45° to the bone wall.

One should distinguish the two types of reinforcing struts; the first one supports the hollow center of the bone against the ovalization, and the second one (an array of crisscrossing struts, with the appearance of a truss) appears at places "in need" supporting the bone against the extensive torsional stresses during the flight (Figure 5a)[8]. Another type of reinforcing structure that helps to withstand the excessive torsional moments of flight are reinforcing ridges (Figure 5b)[9].

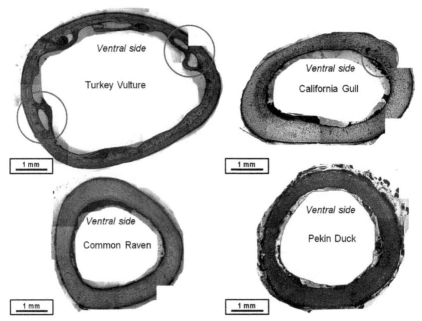

Figure 4. Optical microscopic images of the cross-sections of the ulnae. Struts are shown in red circles.

More detailed analysis of humeri verified that the truss structure is mainly located at the point of pectoral muscle attachment, approximately one-third from the proximal end of humerus at the point of maximum bending and torsional moments carried by the bone in flight[9], as shown in Figure 2. In addition, the struts and ridges were mainly found on ventral side of the humeri and ulnae (Figure 4). Furthermore, the non-flying bird (duck) does not have reinforcing structures; instead both the ulna and humerus have trabecular bone at both proximal and distal ends, similar to mammalian long bones.

Figure 6 shows representative optical images of microstructure of the humeri for all species. The microstructures of the vulture, gull, and raven are similar to fully mature mammalian bone, due to the presence of well-developed microstructural features, such as osteons, Haversian systems, and lacunae, which are observed in the images. The duck humerus is extremely porous and has a less organized structure due to relatively young age of this bird (6 month old) compared to all other bird species (several years old). Reinforcing structures were found in humeri of the flapping/soaring and flapping/gliding birds (Figure 6a and 6b), while humeri of flapping and non-flying birds did not have those structures (Figure 6c and 6d). This observation indicates that reinforcing structures are at the places "in need" that subjected to the maximum stresses during the bird flight.

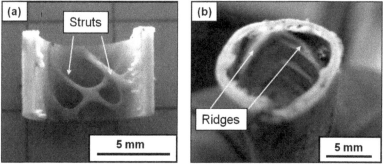

Figure 5. Optical image of (a) reinforcing an array of struts (truss), and (b) reinforcing ridges inside the Turkey Vulture ulna.

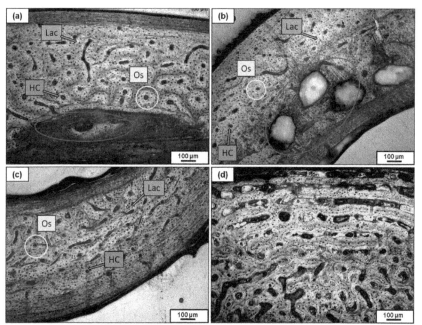

Figure 6. Optical images of humeri of the (a) Turkey Vulture, (b) California Gull, (c) Common Raven and (d) Pekin Duck, showing microstructural features: osteons (Os), lacunae (Lac), and Haversian channels (HC). Ridges are shown in large red ovals.

The data for mineral content, density, and porosity of humeri and ulnae are summarized in Table I. The amount of porosity and mineral content were mostly dependent on the bone maturity level, rather than on the taxa (values for porosity and mineral content were very similar for ulna and humerus of the same species). The amount of porosity for mature birds was found to be between 9-14%. Young duck bones have the highest amount of porosity (~20%) in agreement with our previous study finding young bovine bone to have more porosity than the mature ones[10]. Duck bones also have the lowest mineral content, The density of the humerus was found to be slightly higher than the ulna for flapping/soaring and flapping/gliding birds (vulture and gull), while opposite was found for the flapping and non-flying ones (raven and duck). A possible explanation is that typically gliding and soaring birds have a longer wingspan, resulting in a long moment arm both for torsional and bending moment decomposition of the lift force[9], which translates into a higher stress in the humerus, therefore its density is slightly higher compared to that of ulna. Table I also compared the bird bones with bovine skeletal bone. The bird bones have a smaller amount of minerals and a much larger amount of porosity, which in combination yields a lower density, an advantage for flight.

Table I. Mineral content, density and porosity for humerus and ulna bones of four bird species.

	Mineral content (wt.%)		Density (gm/cm^3)		Porosity (%)	
	Humerus	Ulna	Humerus	Ulna	Humerus	Ulna
Turkey Vulture	60 ± 1	61 ± 2	1.6±0.1	1.2±0.1	11±2	11±2
California Gull	66 ± 1	65 ± 2	1.4±0.1	1.3±0.1	13±3	9±1
Common Raven	64±2	63±1	1.3±0.1	1.5±0.1	14±1	13±3
Pekin Duck	43±1	43±1	1.2±0.2	1.3±0.2	20±4	20±4
Bovine cortical femur bone[14]	65 ± 2		2.0 ± 0.2		8 ± 1	

The Vickers hardness results are summarized in Figure 7. Porosity is one of the main factors contributing to the mechanical properties of bone, along with taxa, hydration condition, anatomical direction, and load distribution[8,13]. Since porosity has adverse effect on strength, the highest porosity of the younger duck bones is in agreement with the smallest hardness values, demonstrating that the mature bone is stronger than the young one. Furthermore, the humerus was found to be significantly harder (p < 0.05) than the ulna for the gull (flapping/gliding bird), while it was opposite for the raven (flapping bird). Potential differences in age (and as a result, diverse microstructure and amount of porosity), as well as different flight behavior (flapping/gliding versus flapping) of these birds are the possible reasons for these results. The hardness values for humerus and ulna were almost the same for the vulture and the duck. These findings again demonstrate that structure and mechanical properties of bird wing bones are optimized for the stresses that those bones are subjected to during bird life. In comparison, the hardness for bovine femur cortical bone is in the range of 550-700 MPa.

To assess the stress distribution in a wing bones subjected to torsion, FEA was applied to the small cylindrical section of the Turkey Vulture ulna (Figure 1b). Two struts with a circular cross-section and diameter equal to the half of the cylinder wall thickness were found in that sample (Figure 1a). The maximum effective strain for the final state of deformation reached a value of 0.13, and no geometrical nonlinearities were developed during the simulation. The von Mises stress distribution obtained by FEA on the top of the bone is shown in Figure 8a, and on the internal bone walls in Figure 8b and 8c. For the given direction of rotation (Figure 1b), the minimum values of von Mises stress were found in the struts and at the immediate areas of their attachment to the bone walls (those areas can be interpreted as a thickening of the cylinder walls). Quantitatively, the struts are subjected to stresses on the order of 15% of those of the inner walls. Due to a larger surface of attachment of the strut in Figure 8c compared to the strut in Figure 8b, the influence of the former one is more notable. These are the preliminary results, and a more detailed modeling of torsional and bending properties for different bird wing bones is been conducted.

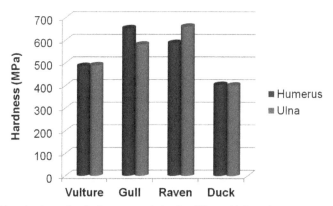

Figure 7. Vickers hardness data for humerus and ulna for different bird species.

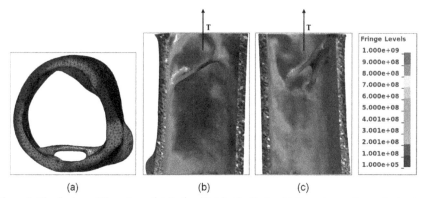

(a)   (b)   (c)

Figure 8. The final von Mises stress distribution [Pa] for the Turkey Vulture ulna section under torsion. (a) Top view, (b) and (c) interiors of two halves.

4. CONCLUSIONS

The structure and mechanical properties of bird humeri and ulnae from flapping/soaring (the Turkey Vulture, *Cathartes aura*), flapping/gliding (the California Gull, *Larus californicus*), flapping (the Common Raven, *Corvus corax,*), and non-flying (the Pekin Duck, *Anas peking*) birds, were investigated by optical microscopy and Vickers hardness testing. The torsional mechanical behavior of a section of the Turkey Vulture ulna was simulated using FEA. The main findings are:

- Wing bones from non-flying birds were found to have a circular cross-section, as well as a uniform thickness around the cross-section due to lack of torsional and bending stresses. In contrast, wing bones from the flying birds experience ovalization and non-uniformity of thickness around the cross-sections due to the large torsional and bending moments during the flight;

- Flying birds have reinforcing structures (struts and ridges) inside their wing bones to optimize and redistribute bending and torsion stresses during the flight;
- Reinforcing struts are roughly at 45° to the bone walls to evenly support the whole structure;
- The humerus was found to be slightly denser compared to ulna for flapping/soaring and flapping/gliding birds, to provide better support for a bird body and redistribute stresses during the flight;
- The detailed geometry measurements by μCT scan together with the capabilities of FEA allowed for an adequate stress distribution determination;
- In the simulation of torsion, the struts and the area around them show the lowest values of von Mises stress on the inner surface of bone under a torsional stress.

ACKNOWLEDGEMENTS

We thank Esther Cory and Professor Robert Sah (UCSD) for the help with μCT imaging, Drs. Cesar Flores and Juan Hermida (Shiley Center for Orthopaedic Research & Education) for their assistance in the a mesh preparation for the FEA, Professor Colin Pennycuick (University of Bristol) and Carlos Ruestes (UCSD) for their valuable insights and discussions. This research was funded by the National Science Foundation, Division of Materials Research, Ceramics Program (DMR 1006931).

REFERENCES

[1]P.-Y. Chen, J. McKittrick, M.A. Meyers, Biological materials: Functional adaptations and bioinspired designs, *Prog. Mater. Sci.,* **57**, 1492-704, (2012).

[2]D.W. Thompson, On Growth and Form, 2nd. Ed. Cambridge University Press, Cambridge, UK, 460, (1968).

[3]A. Wolfson, Recent Studies in Avian Biology, University of Illinois Press, Urbana (1955).

[4]N.S. Proctor, P.J. Lynch, Manual of Ornithology: Avian Structure and Function, Yale University Press, New Haven, USA, (1993).

[5]P.M. O'Connor, L.P.A.M. Claessens, Basic avian pulmonary design and flow-through ventilation in non-avian theropod dinosaurs," *Nature,* **436**, 253-256, (2005).

[6]F.L. Powell, Respiration, in Sturkie's Avian Physiology, edited by Whittow GC, Academic Press, San Diego, 233-264, (2000).

[7]J.D. Currey, Bones: Structure and Mechanics, Princeton University Press, Princeton, NJ, (2002).

[8]B. Bruderer, P. Dieter, A. Boldt, F. Liechti, Wing-beat characteristics of birds recorded with tracking radar and cine camera," *Ibis,* **152**, 272–291, (2010)

[9]C.J. Pennycuick, Modeling the Flying Bird, Elsevier, Oxford, UK, (2007).

[10]Z. Manilay, E.E. Novitskaya, E. Sadovnikov, J. McKittrick, A comparative study of young and mature bovine cortical bone, *Acta Biomater.,* **9**, 5280-5288, (2013).

[11]LS-DYNA Theory manual, (2006).

[12]D. Taylor, J.-H. Dirks, Shape optimization in exoskeletons and endoskeletons: A biomechanics analysis," *J. R. Soc. Interface,* **9**, 3480-3489, (2012).

[13]E. Novitskaya, P.-Y.Chen, E. Hamed, J. Li, V.A. Lubarda, I. Jasiuk, J. McKittrick, Recent advances on the measurement and calculation of the elastic moduli of cortical and trabecular bone: A review," *Theor. Appl. Mech.,* **38**, 209-297, (2011).

[14]E. Novitskaya, A.B. Castro-Ceseña, P.-Y. Chen, S. Lee, G. Hirata, V.A. Lubarda, J. McKittrick, Anisotropy in the compressive mechanical properties of bovine cortical bone: Mineral and protein constituents compared with untreated bone, *Acta Biomater.,* **7**, 3170-3177, (2011).

STRUCTURAL DIFFERENCES BETWEEN ALLIGATOR PIPEHORSE AND BAY PIPEFISH TAILS

Zherrina Manilay[a], Vanessa Nguyen[b], Ekaterina Novitskaya[a,‡], Michael Porter[c], Ana Bertha Castro-Ceseña[a*], Joanna McKittrick[a,c]

[a] Department of Mechanical and Aerospace Engineering, University of California, San Diego, La Jolla, CA 92093, USA
[b] Department of Structural Engineering, University of California, San Diego, La Jolla, CA 92093, USA
[c] Materials Science and Engineering Program, University of California, San Diego, La Jolla, CA 92093, USA

[*] Present address: Instituto Tecnológico de Tijuana, Centro de Graduados e Investigación, Apartado Postal 1166, 22000 Tijuana, Baja California, México.
[‡] Corresponding author, phone: 858-534-5513, fax: 858-534-5698, e-mail: eevdokim@ucsd.edu

ABSTRACT
        This study compares the structure and mechanical properties of the bony plates from the tails of the Alligator pipehorse and the Bay pipefish. These bony plates provide the fish support and protection. The tail structures of both species were investigated by optical and scanning electron microscopy. Partial deproteinization of the samples revealed differences between the connections of the bony plates. The plates of the pipehorse overlap, allowing for flexibility of its tail. In contrast, the plates in the pipefish interlock, making the tail more difficult to bend. These mechanisms contribute to the prehensility and non-prehensility of the tails of the pipehorse and pipefish, respectively.

1. INTRODUCTION
        Advancements in engineering mechanics and sciences often evolve from studying the natural world. Many field of engineering draw inspiration from the observation of biological materials at a high magnification. Several examples of successful engineering innovations inspired by nature are Velcro® inspired by plant burrs[1], adhesive tape inspired by the gecko's feet[2], and the design of wind turbines based on humpback whale tubercles[3-5].
        We previously investigated the structure and unusual deformation mechanisms of the seahorse (*Hippocampus kuda*) tail[6], and now extend the analysis to other members of the family *Syngnathidae*. Two species of pipefish, one with a prehensile tail (Alligator pipehorse, *Syngnathoides biaculeatus*), and one without a prehensile tail (Bay pipefish, *Syngnathus leptorhynchus*), were studied to observe structural differences between the tails. Both species belong to the family *Syngnathidae* along with seahorses and seadragons. Ahnesjö and Craig expressed the significance of the *Syngnathidae* ability to make unusual biological adaptations; recommending this family as a model for studies in evolution[7].
        The Alligator pipehorse has a body that is tapered at both ends; therefore, sometimes it called a "double-ended pipefish." For security and safety, the pipehorse dives vertically down and grasps onto sea grasses for camouflage. The Bay pipefish is also characterized by its slender body shape and camouflages in sea grasses for protection. However, the pipefish does not grasp onto sea grass or swim with a vertical posture. Figure 1 shows photographs of both species of pipefish.
        Both the pipehorse and pipefish are weak swimmers, using their dorsal and pectoral fins as their principal organs of locomotion[8]. Regardless of their similarities, the pipehorse has a prehensile tail that allows it to grasp onto sea grass, while the pipefish lacks prehensility.

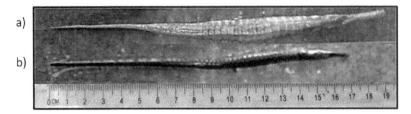

Figure 1. Photographs of (a) Alligator pipehorse and (b) Bay pipefish used in this study.

Similar to the seahorse, the pipehorse and pipefish have external bony plates that are organized along the length of their body[9]. The seahorse has a body covered by ring-like segments made of bony plates connected by rotating and sliding joints that allow it to bend its tail into a logarithmic spiral[6,10-12]. Additionally, these bony plates provide body structure and support. The distinct *S*-shape of the seahorse and its horse-like head distinguishes it from pipefish. Porter et al.[6] found the ring-like segments could withstand large deformations without fracture through local buckling and joint sliding, thereby providing the animal with crushing protection from predators.

There are many articles that discuss seahorses and pipefish from a biological point of view[7-9,13], but little research has been done on the structure of the pipefish armor. Due to the similar habitat and external appearance of both the pipehorse and pipefish, the bony plated structure may be one of the reasons (along with other musculoskeletal components, such as the vertebrae and muscles) of their dissimilar mechanical functionality (prehensility in the former and non-prehensility in the latter). Furthermore, this study may provide a better understanding of the mechanisms that provide prehension in *Syngnathidaes*, which could inspire advancements in mechanical design technologies.

2. MATERIALS AND METHODS

2.1 Sample preparation

A fully grown Alligator pipehorse and a Bay pipefish (see Figure 1) were donated by the Birch Aquarium at Scripps Institute of Oceanography, University of California, San Diego, in June 2012. They died naturally in the aquarium and were kept frozen for several weeks until use. Prior to analysis, the specimens were thawed by immersing them in water at room temperature. Once defrosted, the pipefish were preserved in 70% isopropanol. Several small sections (~ 5 mm length) from the midsection of the tails were cut for analysis.

2.2 Partial deproteinization

The tails were immersed for ~2-3 days in 10-25 ml of an aqueous 0.5 N NaOH solution until the skin was removed and the samples fell apart into individual bony plates[14]. The NaOH solution was replaced daily and sample observations were recorded using an optical microscope.

2.3 Optical microscopy

Partially deproteinized (to better visualize the structure of bony plates and connections) and untreated samples were analyzed by optical microscopy using a Zeiss Axio imager equipped with a CCD camera (Zeiss Microimaging Inc., Thornwood, NY) and a VHX-1000 digital microscope system equipped with a CCD camera (KEYENCE Corporation, Osaka, Japan).

2.4 Micro-computed tomography (μCT) analysis

Whole body samples were investigated by a micro-computed tomography (μCT) scanner, Skyscan 1076 (Kontich, Belgium). The samples were wrapped in a tissue paper that was first moistened with a phosphate buffer saline (PBS) solution and placed in a sealed tube to prevent the specimens from drying out during the scanning process. The samples were scanned using a 0.5 mm aluminum filter. Imaging was performed at 9 μm isotropic voxel size with an electric potential of 48 kV and a current of 200 μA. A beam hardening correction algorithm was used for image reconstruction. Skyscan's Dataviewer and CTVox software were used to analyze and developed images and 3-dimensional (3D) models.

2.5 Scanning electron microscopy

Samples for scanning electron microscopy (SEM) were dried using a critical point dryer (Tousimis Autosanmdri-815, Rockville, MD). The dried samples were sputter-coated with iridium using an Emitech K575X sputter coater (Quorum Technologies Ltd., West Sussex, UK) prior to imaging. The samples were observed at 10 kV with a Philips XL30 field emission environmental scanning electron microscope (ESEM) (FEI-XL30, FEI Company, Hillsboro, OR).

2.6 Microhardness

The microhardness of the untreated bony plates were measured with a micro-hardness testing machine (LECO M-400-H1) equipped with a Vickers indenter. Individual bony plates of each tail segment were embedded in epoxy resin and polished until the surfaces of the samples were exposed. A load of 10 g_f was utilized to indent the exposed surfaces. The Vickers hardness of the bony plates was evaluated by:

$$HV = \left(1.854\frac{F}{d^2}\right) \times 9.81$$

where HV is the Vickers hardness number in MPa, F is the applied load in kg_f, and d is the arithmetic mean of the two measured diagonals in mm.

3. RESULTS AND DISCUSSION

The structural framework of the pipehorse is composed of a vertebral column and bony plates that form ring-like segments. Each segment consists of four bony plates that overlap at the dorsal (back), ventral (front), and lateral (side) mid-lines. The plates decrease in size from the proximal to distal end. At the most distal segments, there are no plates and only the vertebrae remain (Figure 2a). The lack of plates at the distal end is believed to play an important role in the prehensility of Alligator pipehorse tail[15]. μCT imaging in Figure 2a shows the lower part of the pipehorse body, including its prehensile tail. A cross-sectional view (Figure 2b) shows the four bony plates that form the four corners of each segment that surround the central vertebrae.

The pipefish is also composed of a vertebral column and bony plates forming ring-like segments that decrease in size towards the distal end (Figure 2c). Compared to the pipehorse, the main difference is that the plates cover the whole length of the body, and therefore, do not allow prehensility. Figure 2d illustrates that the cross-section of the pipefish tail is also square-shaped, consisting of four bony plates that form each of the four corners. Within each segment, the bony plates overlap at the ventral, dorsal, and lateral mid-lines.

Similar to the pipehorse and pipefish, the bony plates of the seahorse overlap at the mid-section of the dorsal, ventral, and lateral sides of each tail segment[10,16,17]. Hale[10] describes the different joints that connect the bony plates and vertebrae in the seahorse. The transversal and haemal spines (Figures 2b and 2d) connect to the bony plates as pivoting joints, governed by collagenous connective tissues[6,10]. This is similar to a ball-and-socket joint that provides three degrees of rotational freedom[6,10,11]. Within each segment of the seahorse, the bony plates are connected to each other with gliding joints, allowing for translational sliding motion. Gliding joints also connect the segments to one another.

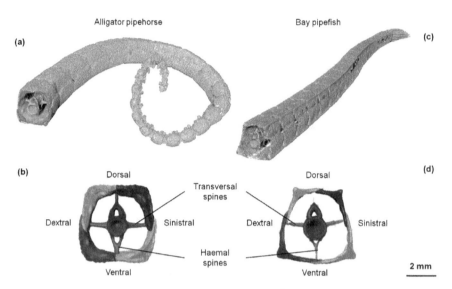

Figure 2. Micro-computed tomography images of the tails and corresponded cross-sections of (a,b) the Alligator pipehorse and (c,d) the Bay pipefish. Colors were added to enhance clarity.

Figure 3 shows SEM images of individual bony plates of the pipehorse and pipefish. The plates of the pipehorse overlap, while the plates of the pipefish interlock (tab and slot). The overlapping area of the pipehorse plates is apparent in the optical micrograph of the midsection of the tail shown in Figure 4. The interlocking joints of the pipefish are shown in the SEM images in Figure 5, showing a tab and slot configuration. This interlocking connection between the bony plates of the pipefish restricts movement between each segment, making it difficult for the tail to bend. In contrast, the overlap of the plates in the pipehorse provides a higher range of mobility between adjacent segments. This helps gives the tail the ability to bend ventrally.

The dorsal views of both tails show connections of the bony plates between each adjacent segment (Figure 6). From these µCT images, it is clear that the bony plates of the pipehorse overlap, allowing for more freedom of movement between each segment, and an overall more flexible tail structure. In contrast, the bony plates between each segment of the pipefish are more restricted, giving the tail a more rigid structure.

Alligator pipehorse   Bay pipefish

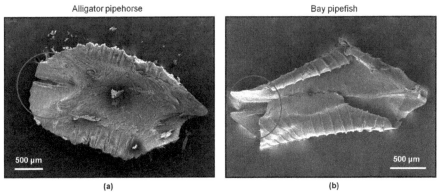

(a)   (b)

Figure 3. SEM images of individual bony plates of (a) Alligator pipehorse, and (b) Bay pipefish (due to extreme thinness, the plate was curled on either side from dehydration). Areas that connect the adjacent segments of the plates are circled in red.

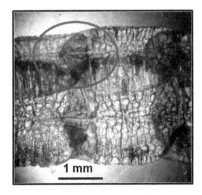

Figure 4. Overlapping bony plates of the Alligator pipehorse under optical microscopy. The section where the two plates overlap is circled.

Microhardness data for both pipehorse and pipefish are summarized in Figure 7. Microhardness testing of the bony plates reveals a similar distribution of hardness across the cross-sections for both species of pipefish that does not seem to vary with location. The average hardness of the pipehorse plates is 400 ± 40 MPa, while that of the pipefish plates is 410 ± 60 MPa. Microhardness data for both species correlate well with microhardness measurements for seahorse bony plates taken with the same load. The average hardness of the seahorse plates was found to be 420 ± 50 MPa[6]. These data indicate that the prehensile ability if the tail does not arise from differences in the mechanical properties of the bone tissue.

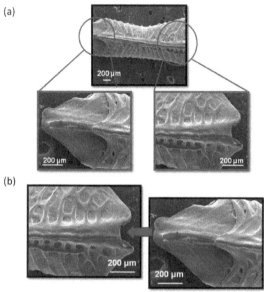

Figure 5. SEM imaging of Bay pipefish bony plates. (a) Top view of a single plate with a magnified view of its ends. (b) Interlocking connection between two plates.

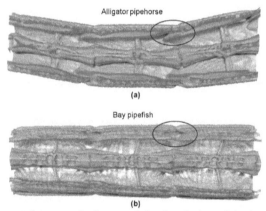

Figure 6. Micro-computed tomography images of the dorsal view of the bony plates of the (a) Alligator pipehorse and (b) Bay pipefish. A segment connection in each tail is circled.

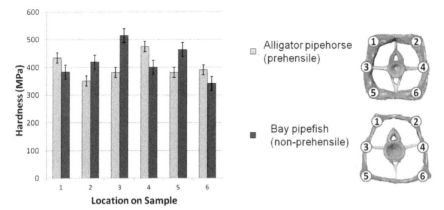

Figure 7. Alligator pipehorse and Bay pipefish hardness values for the cross-sections of the bony plates along with diagrams showing location of indents.

4. CONCLUSIONS

The structure and mechanical properties of the bony plates from the Alligator pipehorse tail and the Bay pipefish tail were analyzed by microscopic methods and hardness testing. It can be concluded that:

- Both species have structural anatomies that are similar, including segments that consist of four bony plates overlapping in the midsection of all four sides;
- The bony plates in adjacent segments overlap in the pipehorse, providing a more flexible structure. In agreement with other researchers[15], the lack of plates at the distal end of its tail provide the pipehorse prehensility;
- The bony plates in adjacent segments interlock in the pipefish, giving it a more rigid structure;
- There is no significant difference between microhardness values of the plates of the prehensile (pipehorse) and non-prehensile (pipefish) tails.

ACKNOWLEDGEMENTS

We thank Leslee Matsushige (Birch Aquarium at the Scripps Institute for Oceanography) for providing the pipefish samples, Ryan Anderson (CalIT2) for his help with scanning electron microscopy, Esther Cory and Professor Robert Sah (UCSD) for the help with μCT imaging. This research is funded by the National Science Foundation, Division of Materials Research, Ceramics Program (Grant 1006931).

REFERENCES
[1]T. Lenau, M. Barfoed, "Material innovation–inspired by nature," *Dan. Metall. Soc.*, **601**, 103-112 (2007).
[2]M. D. Bartlett, A.B. Croll, B.M. Paret, D.J. Irschick, A.J. Crosby, "Looking beyond fibrillar features to scale gecko-like adhesion," *Adv. Mater.*, **24**, 1078–1083 (2012).
[3]E.A. van Nierop, S. Alben, M.P. Brenner, "How bumps on whale flippers delay stall: An aerodynamic model", *Phys. Rev. Lett.*, **100**, 054502-1 – 054502-4 (2008).

[4]F.E. Fish, J.M. Battle, "Hydrodynamic design of the humpback whale flipper," *J. of Morph.*, **225**, 51-60 (1995).

[5]D.S. Miklosovic, M.M. Murray, L.E. Howle, F.E. Fish "Leading-edge tubercles delay stall on humpback whale (*Megaptera novaeangliae*) flippers," *Phys. Fluid.*, **16**, L39-L42, (2004).

[6]M.M. Porter, E. Novitskaya, A.B. Castro-Casena, M.A. Meyers, J. McKittrick, Highly deformable bones: unusual deformation mechanisms of seahorse armor, *Acta Biomater.*, **9**, 6763-6770 (2013).

[7]I. Ahnesjö, J. F. Craig, "The biology of Syngnathidae: Pipefishes, seadragons and seahorses," *J. Fish Biol.*, **78**, 1597–1602 (2011).

[8]M.A. Ashley-Ross, "Mechanical properties of the dorsal fin muscle of seahorse (Hippocampus) and pipefish (Syngnathus)," *J. Exp. Zool.*, **293**, 561-77 (2002).

[9]A.B. Wilson, J.W. Orr, "The evolutionary origins of Syngnathidae: Pipefishes and seahorses," *J. Fish Biol.*, **78**, 1603–1623 (2011).

[10]M.E. Hale, "Functional morphology of ventral tail bending and prehensile abilities of the seahorse, *Hippocampus kuda*," *J. Morphol.*, **227**, 51-65 (1996).

[11]T. Praet, D. Adriaens, S.V. Cauter, B. Masschaele, M.D. Beule, B. Verhegghe, "Inspiration from nature: dynamic modelling of the musculoskeletal structure of the seahorse tail," *Int. J. Numer. Methods. Biomed. Eng.*, **28**, 1028-1042 (2012).

[12]T. Praet, "The biomechanical structure of the seahorse tail as a source of inspiration for industrial design," Ph.D. Thesis, University of Ghent, Ghent, Belgium, http://hdl.handle.net/1854/LU-3262501 (2013).

[13]A.G. Jones, J.C. Avise, "Mating systems and sexual selection in male-pregnant pipefishes and seahorses: Insights from microsatellite-based studies of maternity," *J. Hered.*, **92**, 150-158 (2001).

[14]A.B. Castro-Ceseña, E. E. Novitskaya, P. Y. Chen, G. A. Hirata, and J. McKittrick, "Kinetic studies of the demineralization of bone," *Mat. Sci. Eng.*, **31**, 523-530 (2011).

[15]Professor Dominique Adriaens, private communication.

[16]M.Y. Azzarello, "A comparative study of the developmental osteology of *Syngnathus scovelli* and *Hippocampus zosterae* (Pisces: Syngnathidae) and its phylogenetic implications," *Evol. Monog.*, **12**, 1-90 (1990).

[17]T.N. Gill, "The differential characters of the syngnathid and hippocampid fishes," *Proc. US Natl. Mus.*, **18**,153-159 (1896).

INITIAL INVESTIGATIONS IN APPLYING A PILP-MINERALIZATION SYSTEM TO CALCIUM OXALATE FORMATION USING VAPOR DIFFUSION

Douglas E. Rodriguez, Saeed R. Khan, Laurie B. Gower
Materials Science & Engineering
University of Florida
Gainesville, Florida, USA

ABSTRACT
    In attempts to create an in vitro model system for studying the physiochemical mechanisms involved in the formation of kidney stones, the PILP mineralization process was employed. Here it is hypothesized that the acidic proteins present in urine and renal tissue play a central role in idiopathic nephrolithiasis, where non-classical crystallization may take place. In this two stage process, it has been proposed that calcium phosphate (CaP) is first deposited in the basement membrane of the renal tubules and then grows through the renal interstitium reaching the papillary surface to form sub-epithelial plaque called Randall's plaque (RP). The RP, once exposed to urine in the renal pelvis, becomes coated with calcium oxalate (CaOx) to form a stone. This work is the initial foray into determining the influence of a negatively charged polymer upon the formation of calcium oxalate.

INTRODUCTION
    Pathological biomineralization is a complex process, especially in the formation of kidney stones where the unique and changing urinary environment influences the type and frequency of stone occurrence. While most kidney stones are composed of CaOx and/or some CaP, there is also the presence of an organic matrix composed mainly of acid-rich proteins commonly found in the urine[1, 2]. These urinary macromolecules are thought to modulate mineral precipitation in the urine, which at times is supersaturated with CaOx[3, 4]. The underlying hypothesis of this work is that these macromolecules, although intended to inhibit precipitates, could also play a role in the formation of kidney stones.
    Work in our group has shown that mimics for these acid-rich proteins (such as polyaspartate) can emulate the biomineralization of bone by enabling intrafibrillar mineralization of collagen[5, 6, 7, 8]. In what is termed the polymer-induced liquid-precursor (PILP) process, the presence of negatively charged polypeptides in CaP crystallization solutions results in the formation of an amorphous CaP precursor. The fluidic nature of this precursor allows for infiltration of collagen fibrils, where the precursor then solidifies to amorphous CaP, and then crystallizes to hydroxapatite. It is thought that the acid-rich proteins found in urine, such as osteopontin, could direct an amorphous CaOx precursor in the same way these proteins do in the bone mineralization system.
    Evidence by Evans and coworkers has identified the importance of Randall's plaque in the formation of idiopathic kidney stones[9, 10, 11, 12]. What they observed was that spherules of CaP on the order of 50 nm are deposited in the basement membrane of the kidney. In our studies, we found that mineralization proceeds through collagen supported crystallization of CaP, finally reaching the renal papillary surface and deposition of CaP as sub-epithelial plaques, the Randall's plaques[13]. Exposure of the RP to the urinary environment results in coating with CaOx, where the coating is often in the form of multiple concentric spherical laminations. These CaOx laminations suggest a CaOx amorphous precursor could play a role because we have observed concentric laminations in our PILP formed spherulites of calcium carbonate and phosphate[14, 15, 16]. We found, using fluorescently tagged polyaspartate, that the laminations were a result of exclusion of the polymeric impurity as the amorphous globules crystallized.

This work is a study to identify if polyaspartate can induce the PILP process to result in the formation of a CaOx amorphous precursor. Here, we used a vapor diffusion method to create CaOx using the diffusion of oxalic acid vapors into calcium chloride solutions to induce formation of CaOx. We then studied the effects of the addition of polyaspartic acid (pASP) to the crystallization solution on the subsequent production of CaOx.

METHODS

Calcium oxalate (CaOx) was created by diffusing vapors of oxalic acid into aqueous solutions of calcium chloride. Using a modified approach of Deganello[17], 1 mL of various concentrations of a calcium chloride solution in 50 mM Tris at a pH of 7.5 were added to the center wells of a 24-well cell culture plate, as shown in Figure 1. The $CaCl_2$ concentrations examined were 2 mM, and 10 mM. The wells in the outer columns of the well-plate (refer to Figure 1) each contained 1 mL of deionized water to which 100 µL of diethyl oxalate was added. The remaining wells contained 2 mL of 50 mM Tris. The effects of negatively charged polypeptide were examined by adding either 6.8 kDa pASP (Alamanda Polymers), A) or 14 kDa pASP to the $CaCl_2$ solutions at a concentration of 50 µg/mL.

Cleaned glass cover slips were placed in the wells containing the $CaCl_2$ solution and the well-plates were covered, sealed with parafilm, and placed in a 37 °C oven for 5 days. Cover slips were then removed and rinsed by immersing them in DI water and then ethanol for a few seconds each. After the ethanol evaporated, the cover slips were examined by optical microscopy (Olympus BX60) with an attached polarizer and gypsum 1st-order red wave plate. Raman spectroscopy (Renishaw BioRaman) was also performed in attempts to characterize the mineralization products.

RESULTS

Optical images of the CaOx created in the 2 mM $CaCl_2$ conditions are shown in Figure 2. Here it is seen from Figure 2A that crystals of calcium oxalate dihydrate (COD) were formed with no pASP additive. These crystals range in size from large tetragonal bipyramids to relatively smaller tetragonal bipyramids with flat prism faces. When 6.8 kDa pASP was added to the mineralization solution, longer rod-like tetragonal single crystals of COD formed (Figure 2B). Additionally, small spherical CaOx particles (2 to 5 µm) formed which are thought to be single crystals as indicated by the lack of Maltese crosses in the birefringence pattern. Typical brightfield and polarized images of CaOx formed in the presence of 14 kDa pASP are shown in Figures 2C and 2D, respectively. At first glance the individual crystals appear to be prismatic tetragonal pyramids. Upon closer inspection however, the caps of the crystals have a rough texture, as highlighted by white arrows in Figure 2D. The crystals are also arranged in a spherulitic fashion, although it is difficult to ascertain with these optical images if the crystals are loosely bound to each other or are intergrown together through twinning.

Creation of CaOx at a higher $CaCl_2$ concentration of 10 mM with no pASP again resulted in large tetragonal bipyramids, as shown in Figure 3A. The addition of 6.8 kDa pASP resulted in formation of dumbbell-shaped crystals about 50 µm in diameter (Figures 3B-D), where the open dumbbell seems to have grown until it closes, presumably into a spherulite, as highlighted by the arrowhead in Figure 3D. The dumbbells appear to grow by secondary nucleation or twinning from an individual crystal, as evident from the mid-section of the dumbbell highlighted by the white arrows in Figures 3C and 3D. The addition of pASP again results in smaller spherules, but in this case it cannot be ascertained if the spherules are birefringent. If the spherules do indeed lack birefringence, this is an indication that the spherules have an amorphous structure. The resolution of our optical microscope, however, makes it difficult to ascertain this with certainty.

Similar opened and closed dumbbells (which are spherical in shape) are seen in the 10 mM $CaCl_2$ solution with 14 kDa pASP, as shown in Figures 4A and 4B. These images also show spherules 2 to 4 μm in diameter near the larger dumbbell shaped crystals, as highlighted by arrows. Some of these spherules are birefringent and lack a Maltese cross pattern, which is indicative of single crystals, while it is difficult to see birefringence in other spherules. Figures 4C-4F show higher magnification images of these spherules where slight dimples are observed in some as indicated by arrowheads.

Although SEM was not performed for this particular mineralization experiment, SEM was performed in a previous study in our laboratory for an experiment run under similar conditions (unpublished work). A SEM micrograph of a closed dumbbell crystal from that study is shown in Figure 5A, which appears to be very similar to the closed dumbbells observed with OM in shape, size, and surface texture. Additionally, Raman spectroscopy was performed to further characterize the composition and phase of these dumbbells. The spectrum for the dumbbells in this experiment is shown in Figure 5B, where it is noted that the spectrum is similar to that of COD. Raman was performed on the small (< 5 μm), non-birefringing spherules in attempts to verify the amorphous structure. However, the scans did not yield a suitable signal, presumably because there was not enough of the material gathered in the 10 μm Raman spot.

DISCUSSION

This preliminary study was performed to study the influence of negatively charged polypeptides on the formation of CaOx. The overall goal of this work is to develop a model system to understand the second stage of idiopathic nephrolithiasis, whereby a stone is formed from overgrowth of Randall's plaque with calcium oxalate. It is often thought that the concentric laminations of CaOx found in large stones are caused by sequential layering formed when the urine becomes supersaturated, such as the case of dehydration or hyperoxaluria. An alternative hypothesis is that the formation of these spherical concentric laminations is directed through an amorphous precursor [2], which can lead to diffusion limited exclusion of impurities. This study is focused on the potential generation of an amorphous precursor of CaOx through the use of polyaspartic acid, which has been used extensively by our group to study the PILP process in calcium carbonate. Here, a vapor diffusion set-up was modified where oxalic acid vapors, created through the hydrolysis of diethyl oxalate, diffused into calcium chloride solutions over time. Polyaspartic acid of two different molecular weights (6.8 kDa and 14 kDa) was added to the mineralization solution to examine the effects on formation of CaOx.

The CaOx formed with no pASP additive was in the form of COD, as Deganallo found as well[17]. The COD crystal size here, however, is much smaller than the 80-90 μm crystals they observed, which can be most likely attributed to the much higher $CaCl_2$ concentration in their studies (300 mM versus 2 and 10 mM here). Their study was also conducted at 25 °C versus 37 °C here, so it is expected that there would be an even greater disparity in crystal size than if both studies were conducted at the same temperature.

The addition of pASP resulted in morphologies of large COD crystals (>10 μm) that resembled long, prismatic tetragonal pyramids, often with end caps that showed splitting as in Figure 2D. Some of these larger crystals were also in the form of open and closed dumbbells. These same morphologies were observed in a study examining the effects of polyacylic acid (PAA) on the formation of calcium oxalate, albeit using direct addition crystallization methods [18]. The authors performed atomistic simulations of the interactions of PAA with different crystal faces in attempts to explain the resulting crystal structures. They found that PAA has a preferential association with the {100} face of COD which inhibits crystal growth in the [100] direction, resulting in crystals much longer in the [001] direction, as found in the experiments. Furthermore, they attribute crystal growth into spherulitic structures beginning from crystal branching/splitting from the prismatic faces because of the reduced growth rate of these {100} faces. The branching process then proceeds until the aggregates form dumbbell and sheaf-like morphologies. It stands to reason that this same phenomenon governs the COD crystal growth in this study where pASP should have similar preferential adsorption behavior as PAA, since both are acid rich polypeptides. It is surmised that the rough end caps on the long, prismatic tetragonal pyramids (Figure 2D) is most likely the start of the branching/splitting process. This morphology is limited to the 2 mM $CaCl_2$ condition most likely because there are not enough ions present for crystal growth to proceed in this time period. Conversely, increased kinetics in the 10 mM $CaCl_2$ conditions explain why dumbbell shaped morphologies are observed. Interestingly, no mention of an observed amorphous phase is made in the previous work[19], although because of the size and infrequency of occurrence of the small spherules, it could be easily overlooked or deemed insignificant.

The single-crystalline birefringence of the spherules appears to be very similar to that observed in our experiments with calcium carbonate, where precursor droplets often crystallized into single crystals. Although some amorphous (non-birefringent) CaOx spherules are thought to be observed in this work with polarized OM, further confirmation through Raman spectroscopy was unsuccessful because of an insufficient amount of material. Furthermore, the vapor diffusion method for studying this system may not be the best method for several reasons. First, the CaOx concentration increases slowly over time because of the diffusive nature of the oxalic acid vapor, so early formation conditions will vary from later conditions. Secondly, the pH of the mineralization solution may not necessarily be constant over time because of increasing oxalic acid content, even though the solution was made in 50 mM Tris buffer. Lastly, because the vapor diffuses into the solution at the surface at the top of the fluid, there will be a CaOx concentration gradient in the solution. A direct addition crystallization method may be more suitable to address these issues, and in doing so will be more representative of the urinary environment.

CONCLUSION

The addition of pASP to the CaOx crystallization solution results in the formation of COD morphologies similar to those found by other researchers using PAA in a similar system. Furthermore, while it appears that an amorphous CaOx is formed in this work, further work is needed to confirm this finding. Future studies will most likely be conducted using a direct addition crystallization method in order to better simulate the unique urinary environment.

ACKNOWLEDGEMENT

This work was supported by the National Institutes of Health (grant No. RO1-DK092311). The authors confirm that there are no known conflicts of interest associated with this publication and there has been no significant financial support for this work that could have influenced its outcome.

REFERENCES
[1] S. R. Khan and D. J. Kok, "Modulators of urinary stone formation," *Frontiers in bioscience : a journal and virtual library,* 9 1450-82 (2004).
[2] L. B. Gower, "Biomimetic Model Systems for Investigating the Amorphous Precursor Pathway and Its Role in Biomineralization," *Chemical Reviews,* 108[11] 4551-627 (2008).
[3] S. R. Khan, "Calcium oxalate in biological systems." CRC Press, (1995).
[4] J. A. Wesson, E. M. Worcester, andJ. G. Kleinman, "Role of anionic proteins in kidney stone formation: Interaction between model anionic polypeptides and calcium oxalate crystals," *Journal of Urology,* 163[4] 1343-48 (2000).

[5] T. T. Thula, D. E. Rodriguez, M. H. Lee, L. Pendi, J. Podschun, andL. B. Gower, "In vitro mineralization of dense collagen substrates: A biomimetic approach toward the development of bone-graft materials," *Acta Biomaterialia,* 7[8] 3158-69 (2011).

[6] S.-S. Jee, T. T. Thula, andL. B. Gower, "Development of bone-like composites via the polymer-induced liquid-precursor (PILP) process. Part 1: Influence of polymer molecular weight," *Acta Biomaterialia,* 6[9] 3676-86 (2010).

[7] T. T. Thula, F. Svedlund, D. E. Rodriguez, J. Podschun, L. Pendi, andL. B. Gower, "Mimicking the Nanostructure of Bone: Comparison of Polymeric Process-Directing Agents," *Polymers,* 3[1] 10-35 (2011).

[8] M. J. Olszta, X. Cheng, S. S. Jee, R. Kumar, Y.-Y. Kim, M. J. Kaufman, E. P. Douglas, andL. B. Gower, "Bone structure and formation: A new perspective," *Materials Science & Engineering R-Reports,* 58[3-5] 77-116 (2007).

[9] A. P. Evan, J. E. Lingeman, F. L. Coe, J. H. Parks, S. B. Bledsoe, Y. Z. Shao, A. J. Sommer, R. F. Paterson, R. L. Kuo, andM. Grynpas, "Randall's plaque of patients with nephrolithiasis begins in basement membranes of thin loops of Henle," *Journal of Clinical Investigation,* 111[5] 607-16 (2003).

[10] A. P. Evan, F. L. Coe, J. E. Lingeman, Y. Shao, B. R. Matlaga, S. C. Kim, S. B. Bledsoe, A. J. Sommer, M. Grynpas, C. L. Phillips, andE. M. Worcester, "Renal crystal deposits and histopathology in patients with cystine stones," *Kidney International,* 69[12] 2227-35 (2006).

[11] A. P. Evan, F. Coe, andJ. E. Lingeman, "Response to 'Randall's plaque and cell injury'," *Kidney International,* 71[1] 83-84 (2007).

[12] B. R. Matlaga, F. L. Coe, A. P. Evan, andE. Lingeman, "The role of Randall's plaques in the pathogenesis of calcium stones," *Journal of Urology,* 177[1] 31-38 (2007).

[13] S. R. Khan, D. E. Rodriguez, L. B. Gower, andM. Monga, "Association of Randall Plaque With Collagen Fibers and Membrane Vesicles," *Journal of Urology,* 187[3] 1094-100 (2012).

[14] L. Dai, E. P. Douglas, andL. B. Gower, "Compositional analysis of a polymer-induced liquid-precursor (PILP) amorphous $CaCO_3$ phase," *Journal of Non-Crystalline Solids,* 354[17] 1845-54 (2008).

[15] F. F. Amos, L. Dai, R. Kumar, S. R. Khan, andL. B. Gower, "Mechanism of formation of concentrically laminated spherules: implication to Randall's plaque and stone formation," *Urological Research,* 37[1] 11-17 (2009).

[16] L. Dai, X. Cheng, andL. B. Gower, "Transition Bars during Transformation of an Amorphous Calcium Carbonate Precursor," *Chemistry of Materials,* 20[22] 6917-28 (2008).

[17] S. Deganello, "Growth by vapor diffusion of calcium oxalate dihydrate, calcium oxalate trihydrate and calcite," *Science and technology for cultural heritage,* 1 1-7 (1992).

[18] A. Thomas, E. Rosseeva, O. Hochrein, W. Carrillo-Cabrera, P. Simon, P. Duchstein, D. Zahn, andR. Kniep, "Mimicking the Growth of a Pathologic Biomineral: Shape Development and Structures of Calcium Oxalate Dihydrate in the Presence of Polyacrylic Acid," *Chemistry-a European Journal,* 18[13] 4000-09 (2012).

[19] T. Jung, W. S. Kim, andC. K. Choi, "Crystal structure and morphology control of calcium oxalate using biopolymeric additives in crystallization," *Journal of Crystal Growth,* 279[1-2] 154-62 (2005).

[20] C. G. Kontoyannis, N. C. Bouropoulos, andP. G. Koutsoukos, "Use of Raman spectroscopy for the quantitative analysis of calcium oxalate hydrates: Application for the analysis of urinary stones," *Applied Spectroscopy,* 51[1] 64-67 (1997).

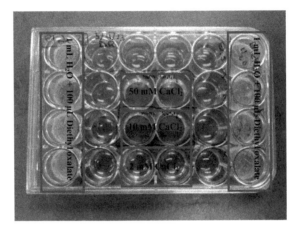

Figure 1. Schematic overlayed on a photograph of a 24-well plate to illustrate the arrangement of solutions in the vapor diffusion set-up. The outer wells contain 1 mL of DI water with 100 μL diethyl oxalate. The middle wells contain 1 mL of 2, 10, and 50 mM $CaCl_2$ made in a 50 mM Tris. The remaining wells contain 2 mL of 50 mM Tris solution.

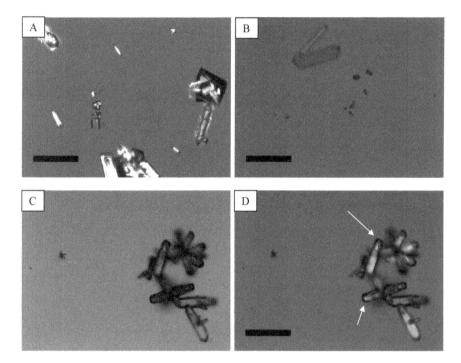

Figure 2. CaOX created in 2 mM CaCl$_2$ conditions with A) no pASP, B) 6.8 kDa pASP, and C) and D) 14 kDa pASP (50 μm scale bar in all images).

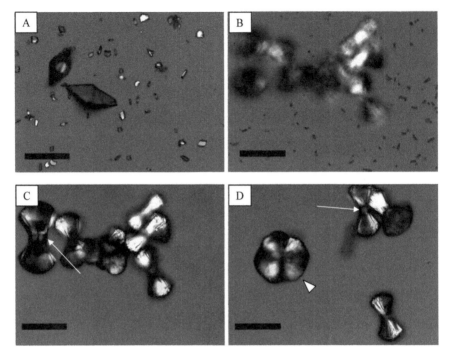

Figure 3. CaOx grown in 10 mM CaCl$_2$ conditions with A) no pASP, and B-D) 6.8 kDa pASP (50 µm scale).

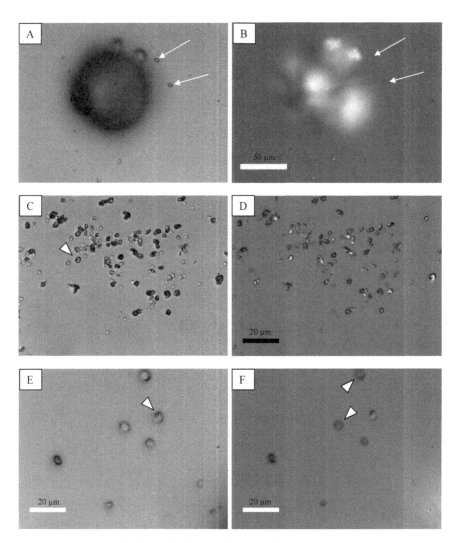

Figure 4. CaOx grown in 10 mM CaCl$_2$ with 14 kDa pASP.

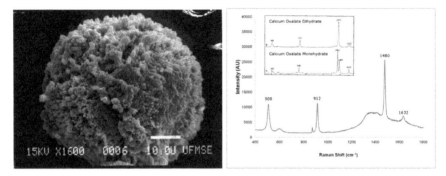

Figure 5. A) SEM micrograph of a closed dumbbell CaOX formed in an earlier vapor diffusion experiment performed in our lab under similar conditions (unpublished work). B) Raman spectra of CaOX dumbbells showing the CaOX is in the form of calcium oxalate dihydrate [Inset is modified from Kontoyannis et al.[20]].

UTILIZING KAOLINITE AND AMORPHOUS CALCIUM CARBONATE PRECURSORS TO SYNTHESIZE ORIENTED ARAGONITIC STRUCTURES

Jong Seto[1,2*]

1 Physical Chemistry, Department of Chemistry, Universität Konstanz, Universitätstrasse 10, 78457 Konstanz Germany

2 Laboratoire Interdisciplinaire sur l'Organisation Nanometrique et Supra Moleculaire, CEA-Saclay, F-91191 Gif-sur-Yvette Cedex, France and Interactions et Dynamique des Environnments de Surface (IDES), Batiment 504, Campus Universitaire d'Orsay F-91405 Orsay Cedex, France

ABSTRACT

Clay materials have built the foundations of modern civilization--improvements to process these materials have quickened their utilization for use in complex load bearing structures. Specifically, with better methods in organizing the constituent metal oxide and organic components in clay, distribution of characteristic nematic as well as smectic phases can be controlled such that defined structures can be constructed from the bottom-up. In this work, we utilize the interactions of an amorphous calcium carbonate (ACC) phase with kaolinite to form a complex composite that can be organized into distinct hierarchical structures. We demonstrate that these ACC - Kaolinite composites can maintain characteristic long range ordered layer-by-layer structures, from the nano- to milli- meters, through convenient and economical processing at room temperature.

INTRODUCTION

Through the use of clay materials, organized load bearing structures can be assembled by bulk processing. These methods have included baking, kilning, as well as air drying–all various forms of dehydration. The dehydration process in clay materials serves two prominent roles in the formation of these materials—removal of a hydration component which primarily enabled a dispersion of reacting elements in solution and the orientation as well as formation of constituent materials into bulk materials that can withstand loading in compression. Furthermore, even after these dehydration processes, the constituent clay materials continue to be reactive and interact with various compounds in the environment due to their innate electrostatic surface behaviors and intermolecular organization. These are the characteristic behaviors whereby many groups have attempted to utilize to capture heavy metal pollutants in aquafiers as well as recently, for uses in carbon capture strategies [1-3].

More specifically, many of these aforementioned clays are composed of silicate components, the most abundant geological mineral on Earth, in either the main tetrahedron or octahedron forms. The silicate constituents are responsible for the short range electrostatic interactions on which orthosilicates, sorosilicates, as well as phyllosilicate based sheets are formed [4]. Specifically, in the tetrahedral form the silicate component is able to form negative surface charges to create hydration layers as well as other binding partners on its surface, enabling for the diverse organization of the silicate subunits found in silicate minerals [5]. Its natural abundance and surface reactivity make silicate-based minerals ideal for creating composite materials [6]. Well known is its application as a binder for cements in the form of calcium silicate hydrate (C-S-H) [7]. In combination with traditional cementious filler materials for concrete, C-S-H is able form a coherent, ordered concrete material for multi-axial load bearing applications. However, the mechanism in which C-S-H is able to order and adhere to concrete filler particles as well as its primary interaction in these cementious materials remains unknown. Another application is the formation of layered polymer composites with smectic clays [8]. It is well established that silicate based materials are invaluable for today's composite and structural materials found in the numerous form of clays and ceramics [4]. Among well-known silicate based clays used in amounts of tons are montmorillonite and kaolinite.

In the case of kaolinite, which has two orders of magnitude reduced cation exchange capacity [CEC] compared to smectic clays like montmorrillonite [9, 10], is a low shrink/swell mineral often used as a filler component [9]. Kaolinite is characterized by its hexagonal plates and phase transformations at high temperatures in atmospheric conditions. Having subunits in the same structural configuration as that of montmorrillonite, it is a layered silicate-alumina-silicate whereby electrostatic interactions are also predominant at the surface [11]. Due to these interactions, kaolinite is a reactive substrate of focus for many groups to bind or modify with other materials in forming unique composite structures—often utilized in blood-clotting as well as paper coating applications [12, 13]. It has been recently established that silica is responsible for stabilization "biomorphic" outgrowths found in geology [14]. In extension to this silica work, silicates possess similar stabilization of amorphous mineral phases traditionally known to be transient and undergoes quick amorphous-to-crystalline transformations [15-17]. In this work, through the novel synthesis and addition of an amorphous calcium carbonate phase at room temperature in the presence kaolinite, we find that the kaolinite is able to serve as a substrate that templates and subsequently, reacts with the ACC phase to induce the crystallization of a preferred aragonitic polymorph of calcium carbonate. We demonstrate that the ACC - Kaolinite interactions occur at short length-scales, but has ramifications for higher length-scales—enabling the possibility for constructing bulk materials that can be aligned and oriented for specific high load bearing applications.

MATERIALS AND METHODS

SYNTEHTIC ACC PRODUCTION

Stable amorphous calcium carbonate was mass produced via fast mixing of a mixture of 2M $CaCl_2$ + 2M $MgCl_2$ and 2M Na_2CO_3 via two HPLC pumps (LC10AS Liquid Chromatograph, Shimadzu Co. Kyoto Japan) connected via a Y-mixer with a 200 µm diameter hole at the junction of the two flowing streams [18]. The mixture consists of nano- to micro- meter sized amorphous calcium carbonate particles with no attention to orientation. This is similar to a technique described by Volkmer and coworkers used to apply thin films as coatings. The ACC was washed several times in ethanol and subsequently, in $CaCO_3$ saturated ddH2O to get rid of the excess sodium chloride. Afterwards, the ACC was spun down in a centrifuge for 5 minutes.

The remaining supernatant was decanted and the ACC fraction was recovered. Using 1 mg/mL of kaolinite argile, the aluminosilicate additive with kaolinite plates in the micrometer size regime was added to the ACC and vigorously shaken to create a homogenous suspension. The resulting mixture was then lyophilized for over 24 hours.

POLARIZED LIGHT MICROSCOPY

An optical microscope (Zeiss Imager M2m, Zeiss GmbH, Jena Germany) with polarized objectives at 5X-100X with a Z-automatic stage was used to analyse the samples. Samples were spread onto a glass slide and imaged in polarized, reflective mode. Z slices at every 100 nm were taken and the slices were averaged over 1 mm to obtain convoluted images. Images were recorded digitally with the manufacturer supplied digital camera (Zeiss mr5c, Zeiss GmbH, Jena Germany).

SCANNING ELECTRON MICROSCOPY (SEM)

Air dried samples at room temperature were placed onto double sided carbon tape for microstructural analysis. A desktop SEM system (Hitachi TM-3000, Hitachi High-Technologies Europe GmbH, Krefeld Germany) at 15 kV was used to examine the morphology of crystalline components. Additional EDX analyses were performed with a solid state EDX detector (Xcite dectector, Bruker AXS GmbH, Berlin Germany) attached to the SEM setup.

TRANSMISISON ELECTRON MICROSCOPY (TEM)

Samples were aliquoted onto 400 mesh copper grids coated with carbon film (Quantifoil Micro Tools GmbH, Jena Germany) and allowed to air dry under clean room conditions at the Nanostructure Laboratory of the University of Konstanz. The sample grids were imaged with an in lens column filter TEM (Zeiss Libra 120, Zeiss SMT GmbH, Oberkochen Germany) at 120 mV with 1 mrad at magnification series (8-100 kX) to examine micro- and sub- structures of each sample.

ATOMIC FORCE MICRSCOPY (AFM) IMAGING

An AFM imaging setup (Nanowizard, JPK Instruments AG, Berlin, Germany) with silicon nitrade AFM tips (Olympus Corp., Tokyo, Japan) was used to scan sample surfaces in contact mode. Using scan rates of 10 microns/second and in phase imaging mode, 100 X 100, 50 X 50, and 30 X 30 micron areas of the samples were imaged.

ATTENUATED TOTAL REFLECTANCE- FOURRIER TRANSFORM INFRARED (ATR-FTIR) SPECTROSCOPY

A standard ATR-FTIR spectrophotometer (Spectrum 100, Perkin-Elmer Life and Analytical Sciences, Bridgeport, CT USA) was used to obtain spectra between 4000-650 cm^{-1} wavenumbers for each sample. The spectra were analyzed with a spectrum analysis program (Spectrum version 6.2.0, Perkin-Elmer Life and Analytical Sciences, Bridgeport, CT USA) supplied by the FTIR manufacturer.

X-RAY DIFFRACTION (XRD)

A powder X-ray diffractometer (D8 Advance, Bruker AXS GmbH, Berlin Germany) was used to measure the crystalline reflections of the samples with a scan time of 1 hour each and a scan range $2\theta \sim 10° - 80°$.

SMALL-ANGLE X-RAY SCATTERING (SAXS) SPECTROSCOPY

A lab source SAXS/WAXS system (Nanostar, Bruker AXS GmbH, Berlin Germany) with a copper anode generator operated at 40 kV and 35 mA measured samples for 1 hour. Subsequent, data analysis and integrations (both radial and azimuthal) of the SAXS patterns were accomplished using Fit2D (AP Hammersley, *ESRF Internal Report*, ESRF98HA01T, FIT2D V9.129 Reference Manual V3.1 (1998)).

THERMOGRAVIMETRIC ANALYSIS (TGA) and DIFFERENTIAL THERMAL ANALYSIS (DTA)

Approximately 25-30 mg of sample were weighed and used for thermoanalyses in a standard thermogravimetric analyzer (STA 429, Netzsche GmbH, Selb Germany) over a temperature range of 25°C - 950°C. Both TGA and DTA analyses of all samples were measured under O_2.

RESULTS AND DISCUSSION

Crystalline kaolinite plates have the capability to assemble along specific planes due to weak electrostatic and hydrophilic forces found on their [001] surfaces [12]. Several groups have implicated the assembly process as a result of surface charge density, suggesting that the charge densities of kaolinite plates are highly dependent on pH [19] [20-22]. In a similar fashion, amorphous calcium carbonate (ACC) are attracted to these same kaolinite surfaces via cation bridging due to their high charge and hydration densities, enabling the diffusion and adsorption of ACC onto the kaolinite surfaces to occur more quickly than the assembly of kaolinite-kaolinite plates. Thus, the ACC - Kaolinite interaction inadvertently, mediates an

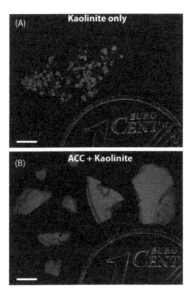

Figure 1. Macroscopic comparison of the packing of kaolinite and ACC + Kaolinite hybrid composites (A.) Kaolinite only (B.) ACC + Kaolinite [scale bar = 2 mm]

alternating layer-by-layer structure of ACC and kaolinite. As observed under polarized light microscopy, kaolinite platelets assemble with immediately adjacent platelets and demonstrate partial aggregation as well as a degree of polydispersity. In contrast to this kaolinite platelet only scenario, the addition of ACC to kaolinite platelets enable aggregation into larger aggregate species (> 100 microns) and the constituent kaolinite and ACC are organized along specific orientations (Figure 1). This orientation can be qualitatively observed under polarized light microscopy (Figure 2).

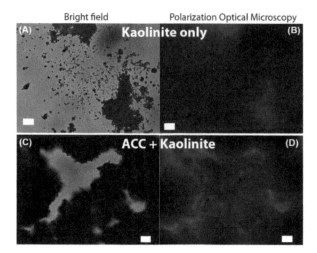

Figure 2. Polarized optical light microscopy of kaolinite only and amorphous calcium carbonate (ACC) + Kaolinite hybrid composite samples (A.) Bright field of kaolinite only sample (B.) Polarized microscopy of kaolinite only sample (C.) ACC + Kaolinite sample in brightfield (D.) ACC + Kaolinite sample under polarized microscopy [scale bars = 20 μm]

By examining these samples more closely under scanning electron microscopy (SEM), it can be confirmed that the kaolinite only samples demonstrate a short range order such that stacks of immediate neighboring kaolinite plates accrue together. This is in contrast to the ACC - Kaolinite composite fractions where aggregates show orientation and assembly on a larger length-scale (Figure 3). In comparison to kaolinite only aggregation, the ACC - Kaolinite

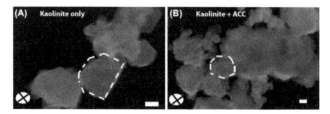

Figure 3. Scanning electron microscopy of kaolinite and ACC + Kaolinite hybrid composite samples (A.) Individual kaolinite platelets can be seen to be clustered together (B.) In the presence of ACC, kaolinite platelets appear to aggregate such that individual platelets can no longer be delineated [scale bars = 200 nm]; dotted lines delineate kaolinite border and dot with "X" indicates relative orientation of the plates

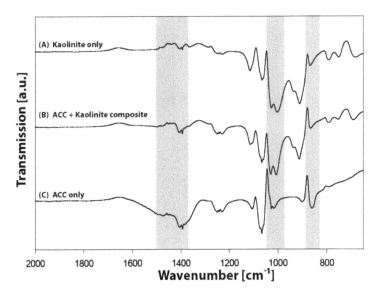

Figure 4. ATR-FTIR spectroscopy of ACC-Kaolinite hybrid composite with ACC $\gamma 4$, $\gamma 2$, $\gamma 3$ regions highlighted in grey (860-866 cm-1 , 1000-1050 cm-1, and 1400-1480 cm-1, respectively) (A.) Kaolinite only (B.) ACC - Kaolinite composite (C.) ACC only

samples demonstrate stacks which are thicker and cover a longer length-scale. This observation implicates the ACC component as a surface functionalizing agent such that adjacent kaolinite platelets are more prone to aggregate together to result in the formation of larger species. Interestingly, the ACC found at the inter-platelet regions cannot be detected by techniques like FTIR and SAXS due to the thickness, only ACC found at the surface of these hybrid composites are detected (Figure 4 and Figure 5). In FTIR, contribution of ACC to the kaolinite can be witnessed from the broadening of the $\gamma 4$ and $\gamma 2$ peaks, the in-plane bending and out-of-plane vibration, respectively, suggesting the presence of a disordered phase. Typically, the ratio of the $\gamma 4$ and $\gamma 2$ peaks has been used to determine ACC content in FTIR samples. From Figure 3a, each platelet of kaolinite can be well defined and delineated. The platelets are approximately 100-400 nm in diameter and appear to be pseudo-hexagonal in morphology. In contrast, Figure 2b shows ACC - Kaolinite species to be 100 nm – 2 μm in diameter and assume a plate-like clustering along the [001] face of each plate. This structuring at lower length-scales lends itself to further organization at the macroscale (Figure 1).

From transmission electron microscopy (TEM), the stacking of the kaolinite platelets occur together such that each stack contains 2-4 platelets in the kaolinite only situation (Figure 3a) with each platelet being ~10s of nm thin due to the penetration of the electron beam and previously reported in literature [11]. Unlike the kaolinite only samples, the ACC - Kaolinite

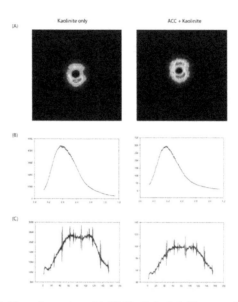

Figure 5. Small-Angle X-ray Scattering of ACC-Kaolinite hybrid composite (A.) SAXS patterns (B.) Radial integration (C.) Azimuthal integration

samples is observed to stack similarly such that the platelets are organized parallel to <001>, but are also laterally organized such that neighboring stacks of platelets form a uniform super-aggregate (Figure 3b). From the electron diffraction (ED) patterns, the kaolinite only samples can be seen to display the polydispersity observed in polarized light microscopy. Interestingly, this polydispersity can also be witnessed in the ACC - Kaolinite aggregates suggesting that the kaolinite constituent forms first and further assembly takes place in the presence of ACC. In the ED pattern of the ACC - Kaolinite sample, in addition to an amorphous halo indicating the existence of an ACC phase, the formation of an aragonitic structure can be observed and assigned including the [001] zone axis found typically in aragonitic structures (Figure 6f). PXRD also confirms the existence of an aragonite species in the ACC – Kaolinite composite (Figure 7). The kaolinite crystalline structure does not match an aragonite {001}, ruling out the possibility that the kaolinite surface templates an aragonitic structure [21] [23]. From previous work, many groups have found that high amounts of Mg^{2+} can induce aragonite formation; where significant amounts of Mg^{2+} can be found along with the kaolinite particles. The Mg^{2+} content may compete with the Ca^{2+} such that a mix of Ca^{2+} and Mg^{2+} in a specific ratio will be found at the surface of the kaolinite platelets. This ratio of Ca^{2+} and Mg^{2+} will induce a specific lattice ordering when mineralization occurs—favoring aragonite. Thus, this aragonite most likely arises from the excess amount of Mg^{2+} used in the stabilization of the ACC phase when forming the ACC - Kaolinite hybrid composite. The excess Mg^{2+} subsequently causes the ACC to undergo crystalline transformation to aragonite [24] [25] [26] [27]. In Figure 6, when examining the ACC fraction without the presence of any additives, it can be observed

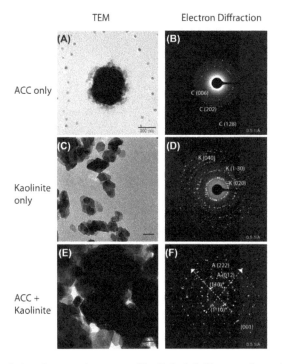

Figure 6. Transmission electron microscopy of kaolinite hybrid composites and corresponding electron diffraction patterns (A.) ACC only (B.) ED pattern of ACC displaying diffraction spots of calcite and vaterite (C.) Kaolinite platelets (D.) ED pattern of kaolinite (E.) ACC + Kaolinite composite (F.) ED pattern of ACC + Kaolinite displaying two diffraction spots, the zone axis of [001], and the twinning places of {110} of aragonite

that the hydrated species slowly transforms into crystalline species, specifically calcite and vaterite. From the TEM and PXRD observations, we demonstrate that the ACC – Kaolinite hybrid composite undergoes an ACC-crystalline phase transformation such that a predominant aragonitic phase is formed.

This stacking phenomena observed in the kaolinite only samples is most likely explained by the weak electrostatic forces on the [001] basal and [010] edge surfaces of each kaolinite platelet produced typically by an Al(III) substitution for Si(IV) in the silicate layer [11, 20]. Zeta potential measurements through various methods such as electrophoresis, electroacoustic, and titration have found that these quantities are inaccurate due to inhomogenous charge densities as a result of the edge surface [19] [11, 20] [28, 29]. Instead, using scanning probe microscopy measurements on kaolinite plate surfaces at varying pH, it was determined that the surfaces

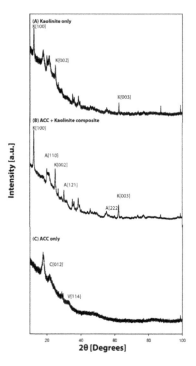

Figure 7. Powder X-ray diffraction of ACC-Kaolinite hybrid composites (A.) Kaolinite only (B.) ACC + Kaolinite composite (C.) ACC only

potentials and charge densities dramatically decrease with increasing pH, possibly explaining the predominant attractive forces between kaolinite plates observed at high pHs [21] [22] [23]. Having less H^+ will induce cation bridging across the charged kaolinite plates, while having more H^+ will saturate the surfaces of the platelets and disrupt the native aggregation characteristics of the kaolinite plates. From the native charge densities on the kaolinite plate surfaces and through weak counter ions like Ca^{2+} and Mg^{2+} found in solution mediating the interaction, the platelets of kaolinite can accrue via a "platelet-by-platelet" fashion. Through the limited accumulation of these counter ions on the surface, long range stacking is limited. Furthermore, these interactions are highly pH sensitive and due to fluctuations in solution pH, the charge densities can also vary and result in various assemblies [11, 22]. By measuring the solution pH of the forming ACC and Kaolinite composite, it was found that the ACC – Kaolinite hybrid composites assemble under a pH of 10.1 environment, suggesting that kaolinite-kaolinite plate stacking is the initially preferred orientation in the presence of ACC. This can be confirmed in both SEM and TEM, whereby longer range aggregation and stacking can be observed when an ACC component is added to kaolinite (Figure 6). Specifically, the ACC is able to interact with the [001] surface of each kaolinite surface along a similar cation ion bridging mechanism found in the kaolinite only

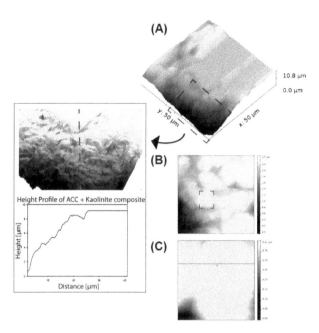

Figure 8. Atomic Force Microscopy (AFM) of ACC - Kaolinite hybrid composites (A.) Height profile of ACC + Kaolinite composite with subsequently higher magnifications (B.) 10 x 10 μm area scan of (A) (C.) 1.5 x 1.5 μm area scan of the region of interest highlighted in (B)

p stacking. In contrast, the ACC fraction contains high amounts of Ca^{2+} and Mg^{2+} ions, allowing bridging across kaolinite platelets over longer length-scales (Figure 8). The ACC at the inter-late spaces subsequently dehydrates and in the presence of Mg^{2+}, eventually proceeds to undergo an amorphous-to-crystalline transformation into aragonite [24, 25, 30] [27]. An aragonitic polymorph is preferentially formed instead of the more stable calcite polymorph due to the divalent Mg^{2+} substitution of Ca^{2+} [26]. This layer-by-layer aggregation of ACC and kaolinite enables for long range ordered structures as observed frequently in biominerals whereby an organic template is able to direct the formation of an aragonitic plate formation.

From further analyses via AFM, we observe that organization of the ACC - Kaolinite composite stacks is indeed in the direction parallel to the [001] direction. Additionally, we find that the lateral interactions of the ACC - Kaolinite species allows for convergence of two neighboring stacks into one super-stack, enabling for seamless joining of stacks —hence, the long ordering of Kaolinite plates. This enables for a step-like growth plateau observed in the macrostructure of the ACC - Kaolinite hybrid composites (Figure 8 inset). Specifically, ACC is able to act as a molecular binder such that kaolinite-kaolinite repulsion is reduced and mediates a growth layer where other ACC and kaolinite components can interact and further assemble seamlessly into higher ordered structures (Figure 8). These observations reveal that ACC itself

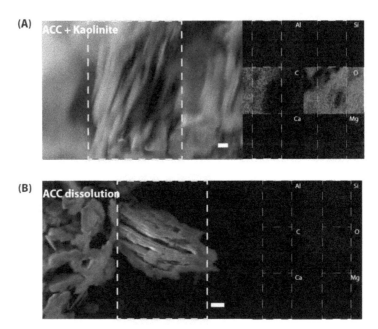

Figure 9. Elemental distribution of ACC - Kaolinite hybrid composites (A.) End on view of a stack of kaolinite plates in the presence of ACC and elemental maps of the stacks (B.) End on view of a stack of kaolinite plates after dissolution of the ACC component of a ACC – Kaolinite sample [scale bar = 200 nm]

possesses the ability to fill in small spaces possibly reducing interfacial energies between kaolinite-kaolinite platelets, acting similarly to a molecular cement. This stabilization by ACC is significant in constructing larger kaolinite-related structures due to the surface charge forces becoming extremely weak over large length-scales [31]. An ACC component can be observed to be distributed throughout all surfaces of the ACC – Kaolinite composite (Figure 9a). Without an ACC component, kaolinite-kaolinite assembly does not occur beyond 400 nm, most likely due to the weakening of these aforementioned surface charge forces at this length-scale (Figure 8 b,c). Interestingly, when the ACC is removed through dissolution by water over time, the remaining stacked kaolinite platelets retain the long range ordered arrangements (Figure 9b). Through this stabilization mechanism by the ACC component, ACC - Kaolinite can stack into structures with longer length-scales.

We observe an ACC - Kaolinite composite material that self assembles into long range ordered layered composite through shorter range interactions. By being able to assemble at short length-scales, ACC constituents can be packed into the inter-platelet spaces of kaolinite.

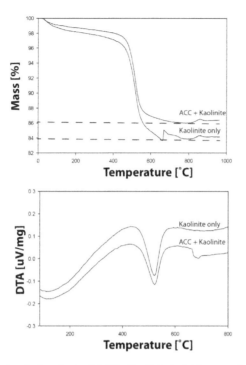

Figure 10. Thermochemical analysis of ACC - Kaolinite hybrid composite (A.) TGA scan from 0-1000 °C.(B.) DTA scan from 0-800 °C

However, the mechanism in which the ACC and Kaolinite interacts to form an aragonite – Kaolinite hybrid composite is still not entirely clear. We do know from TGA the contribution of the ACC to the mass of the entire hybrid composite is ~2% (Figure 10.) And through interaction of the ACC onto the surfaces of kaolinite platelets as well as filling the inter-platelet spaces via cation bridging, the ACC is found to transform to an aragonitic polymorph. Even with this aragonitic crystalline component, the ACC – Kaolinite composite still consists of a large portion of an amorphous phase (Figure 9). This is the first report of an inorganic template for aragonite formation without the use of any organic additives as aragonite formation and stabilization has been the domain of biologically derived minerals [32] [33] [34] [35]. A schematic of this ACC - Kaolinite assemblage is shown in Figure 11.

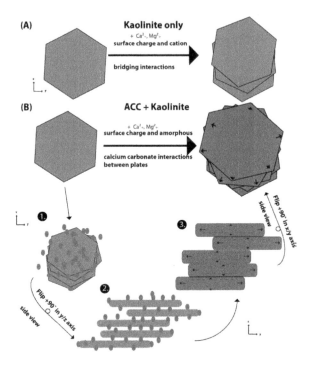

Figure 11. Schematic of kaolinite and ACC + Kaolinite hybrid composite stack assembly (A.) Kaolinite organization is dependent on cation bridging interactions between other kaolinite platelets (B.) ACC is intercalated into the interplatelet space of kaolinite and increases the interactions at the platelet interface according to the following steps: 1. Attraction and adsorption of ACC onto kaolinite, 2. Adsorption of ACC into the inter-platelet spaces, 3. Aggregation and filling in of residual ACC on to kaolinite surfaces via cation bridging

CONCLUSIONS

We demonstrate here that due to the synergistic interactions of ACC and kaolinite, at the molecular scale, an ordered hierarchical composite is formed that spans at length-scales larger than 1 mm. We find that the ACC forms a layer around the kaolinite platelets and enable for binding of adjacent ACC - Kaolinite hybrid platelets such that long range ordering in the x/z and z/y directions to create large ACC - Kaolinite assemblies. Interestingly, with further modifications such as the monolayer addition of known organic components with specific catalytic or photothermal conversion properties, we can enable this ACC - Kaolinite hybrid composite to be a model system to understand the effect of specific molecules on structure and function of a hierarchically, self-assembled composite—perhaps not only endowing this composite with similar bulk material properties found in the biological world [36] [37], but properties with tunable organization and mechanical performance.

ACKNOWLEDGEMENTS

The author would like to thank his family and Reena Bajwa for their continued support and enthusiasm in work towards sustainable processing of bioinspired materials. J.S. would also like to acknowledge Ulrich Tritschler and Marius Schmidt for their technical assistance in preparing the ACC – Kaolinite composite samples. Andreas Picker is acknowledged for his fruitful discussions on the manuscript.

REFERENCES

1.   Choi, S., et al., *Strontium speciation during reaction of kaolinite with simulated tank-waste leachate: Bulk and microfocused EXAFS analysis* Environmental Science and Technology, 2006. **40**(8): p. 2608-2614.
2.   Karoui, H., et al., *Influence of clay suspensions on the precipitation of CaCO3 in seawater* Crystal Research and Technology, 2010. **45**(3): p. 259-266.
3.   Behara, S.K., *Sorptive removal of ibuprofen from water using selected soil minerals and activated carbon.* International journal of environmental science and technology, 2012. **9**(1): p. 1735-1472.
4.   Liebau, F., *Structural Chemistry of Silicates*1985: Springer-Verlag. 347.
5.   Knight, C.T.G., R.J. Balec, and S.D. Kinrade, *The Structure of Silicate Anions in Aqueous Alkaline Solutions.* Angewandte Chemie International Edition, 2007. **46**: p. 8148-8152.
6.   Kleinfeld, E.R. and G.S. Ferguson, *Healing of defects in the stepwise formation of polymer/silicate multilayer films.* Chem. Mater., 1996. **8**(8): p. 1575-1778.
7.   Richardson, I.G., *The calcium silicate hydrates.* Cement and Concrete Research, 2008. **38**(2): p. 137-158.
8.   Alexandre, M. and P. Dubois, *Polymer-layered silicate nanocomposites: preparation, properties and uses of a new class of materials.* Materials Science and Engineering R, 2000. **28**(1-2): p. 1-63.
9.   Carroll, D., *Ion exchange in clays and other minerals.* Geological Society of America Bulletin, 1959. **70**(6): p. 749-780.
10.   Mengel, D.D., *Fundamentals of Soil Cation Exchange Capacity*, 1993, Department of Agronomy, Purdue University: Indianapolis.
11.   Zhou, Z. and W.D. Gunter, *The Nature of the Surface Charge of Kaolinite.* Clays and Clay Minerals, 1992. **40**(3): p. 365-368.
12.   Murray, H.H. and S.C. Lyons, *Correlation of paper-coating quality with degree of crystal perfection of kaolinite.* Clays Clay Mineral, 1956. **456**: p. 31-40.
13.   Baker, S.E., et al., *Blood Clot Initiation by Mesocellular Foams: Dependence on Nanopore Size and Enzyme Immobilization.* Langmuir, 2008. **24**(24): p. 14254-14260.
14.   Kellermeier, M., et al., *Evolution and Control of Complex Curved Form in Simple Inorganic Precipitation Systems.* Cryst. Growth Des., 2012. **12**(7): p. 3647-3655.

15. Kellermeier, M., et al., *Stabilization of Amorphous Calcium Carbonate in Inorganic Silica-Rich Environments.* J. Am. Chem. Soc., 2010. **132**(50): p. 17859-17866.
16. Kralj, D. and N. Vdovic, *The influence of some naturally occurring minerals on the precipitation of calcium carbonate polymorphs* Water Research, 1999. **34**(1): p. 179-184.
17. Lee, B.J., *Competition between kaolinite flocculation and stabilization in divalent cation solutions dosed with anionic polyacrylamides.* Water Research, 2012. **46**(17): p. 1043-1354.
18. Malinova, K., et al., *Production of CaCO3/hyperbranched polyglycidol hybrid films using spray-coating technique.* J Colloid Interface Sci., 2012. **374**(1): p. 61-69.
19. Ferris, A.P. and W.B. Jepsen, *The exchange capacities of kaolinite and the preparation of homoionic clays.* J Colloid Interface Sci., 1975. **51**(2): p. 245-259.
20. Schroth, B.K. and G. Sposito, *Surface Charge Properties of Kaolinite.* Clays and Clay Minerals, 1997. **45**(1): p. 85-91.
21. Gupta, V., et al., *Crystal lattice imaging of the silica and alumina faces of kaolinite using atomic force microscopy.* J Colloid Interface Sci., 2010. **352**(1): p. 75-80.
22. Gupta, V., et al., *Particle interactions in kaolinite suspensions and corresponding aggregate structures.* J Colloid Interface Sci., 2011. **359**(1): p. 95-103.
23. Yin, X.H., et al., *Surface charge and wetting characteristics of layered silicate minerals* Advances in Colloid and Interface Science, 2012. **179**(SI): p. 43-50.
24. Sugawara, A. and T. Kato, *Aragonite CaCO3 thin-film formation by cooperation of Mg2+ and organic polymer matrices.* Chem. Comm., 2000. **6**: p. 487-488.
25. Huang, Y.C., et al., *Calcium-43 NMR studies of polymorphic transition of calcite to aragonite.* J Phys Chem B, 2012. **116**(49): p. 14295-14301.
26. Menadakis, M., G. Maroulis, and P.G. Koutsoukos, *Incorporation of Mg2+, Sr2+, Ba2+ and Zn2+ into aragonite and comparison with calcite.* J Math Chem, 2009. **46**: p. 484-491.
27. Loste, E., et al., *The role of magnesium in stabilizing amorphous calcium carbonate and controlling calcite morphologies.* J Crystal Growth, 2003. **254**: p. 206-218.
28. Tombacz, E. and M. Szekeres, *Surface charge heterogeneity of kaolinite in aqueous suspension in comparison with montmorillonite.* Appl. Clay Sci., 2006. **34**: p. 105-124.
29. Tribe, L., R. Hinrichs, and J.D. Kubicki, *Adsorption of Nitrate on Kaolinite Surfaces: A Theoretical Study* Journal of Physical Chemistry, 2012. **116**(36): p. 11266-11273.
30. Heywood, B.R. and S. Mann, *Molecular Construction of Oriented Inorganic Materials: Controlled Nucleation of Calcite and Aragonite under Compressed Langmuir Monolayers.* Chem. Mater., 1994. **6**(3): p. 311-318.
31. Israelachvili, J., *Intermolecular and Surface Forces* 2010: Academic Press. 674.
32. Falini, G., et al., *Control of aragonite or calcite polymorphism by mollusk shell macromolecules.* Science, 1996. **271**(5245): p. 67-69.
33. Belcher, A.M., et al., *Control of crystal phase switching and orientation by soluble proteins.* Nature, 1996. **381**: p. 56-58.
34. Liu, F., et al., *Biomimetic fabrication of pseudohexagonal aragonite tablets through a temperature-varying approach.* Chem. Comm., 2010. **46**: p. 4607-4609.
35. Beniash, E., et al., *Amorphous calcium carbonate transforms into calcite during sea urchin larval spicule growth.* Proc. Roy. Soc. B, 1997. **264**: p. 461-465.
36. Nassif, N., et al., *Amorphous layer around aragonite platelets in nacre.* PNAS, 2005. **102**(36): p. 12653-12655.
37. Seto, J., et al., *Structure-property relationships of a biological mesocrystal in the adult sea urchin spine.* PNAS, 2012. **109**(10): p. 3699-3704.

USE OF BIOMINERALIZATION MEDIA IN BIOMIMETIC SYNTHESIS OF HARD TISSUE SUBSTITUTES

A. Cuneyt Tas
Department of Materials Science and Engineering, University of Illinois
Urbana, Illinois 61801, USA

ABSTRACT
    Blood is the medium which provides the vital inorganic ions, biological molecules, growth factors, proteins, and vitamins to the nanocrystals formed by cells in human hard tissues, i.e., bones and teeth. Blood pH is stabilized at 7.4 by a fine balance between carbonic anhydrase and bicarbonate ions. For a healthy person, blood temperature is confined to 36.5°C. Inorganic ions (such as Ca^{2+}, Mg^{2+}, Na^+, K^+, HCO_3^-, HPO_4^{2-}, Cl^-) present in blood are also found in the nanocrystals of hard tissues. These nanocrystals are not stoichiometric hydroxyapatite, but they are non-stoichiometric and heavily doped substances. Biomimetic syntheses performed at 36.5°C and pH 7.4 in synthetic biomineralization media, free of any synthetic, man-made organics, produce calcium phosphates of high BET surface area. This article focuses on summarizing our continuing efforts in developing new biomineralization media (BM-3, BM-7, Lac-SBF, Tris-SBF and urea-enzyme urease-buffered solutions) and the biomimetic synthesis of non-stoichiometric and doped calcium phosphate- and calcium carbonate-based hard tissue substitute materials. The specifics of media development and the full characterization (via electron microscopy, XRD, FTIR, ICP-AES, BET, and cell culture) of the synthesized biomaterials are summarized.

BIOMIMETIC SYNTHESIS

    If CO_2 gas is bubbled through an aqueous solution of 2.5 mM $CaCl_2 \cdot H_2O$, it will start precipitating sub-micron particles of $CaCO_3$ (either calcite or a biphasic mixture of calcite and vaterite depending on the solution pH, CO_2 bubbling rate, and the solution temperature and stirring rate). Human blood has a Ca^{2+} concentration of 2.5 mM. CO_2 bubbles first cause the formation of carbonic acid, H_2CO_3, which then dissociate into HCO_3^- and CO_3^{2-}. The reversal of these dissociation reactions will cause the release of CO_2 gas out of the solution, and a slight increase in solution pH. Marques et al. [1] have explained the buffering effect of HCO_3^-/CO_3^{2-} pair. However, an addition of 27 mM $NaHCO_3$ to the above solution can eliminate the necessity of bubbling CO_2, and it would again form $CaCO_3$ precipitates. Human blood has a HCO_3^- concentration of 27 mM. Then, if one adds 1 mM Na_2HPO_4 into this solution, it will precipitate calcium phosphates instead of $CaCO_3$. Human blood has a P concentration of 1 mM. This solution has a Ca/P molar ratio of 2.50, just like the human blood. This solution has an ionic strength of 75 mM. It is saturated with respect to the formation of apatitic calcium phosphates, and it contains Ca^{2+}, HPO_4^{2-}, HCO_3^-, Na^+, and Cl^- ions. When the pH of the above solution is brought down to the range of 5 to 6, for instance, by adding tiny droplets of dilute HCl, it will only precipitate $CaHPO_4 \cdot 2H_2O$ (brushite). If the pH value of that solution is kept at the physiological value of 7.4 or higher, it would precipitate carbonated, apatitic calcium phosphates. Over the pH range of 6.2 to 10, HCO_3^- is the most stable carbonate species in aqueous solutions. The precipitation of carbonated, Ca-deficient and apatite-like (apatitic) calcium phosphates from such a solution is described by the following reaction:

$$(10\text{-}x\text{-}y)Ca^{2+} + (6\text{-}x\text{-}y)HPO_4^{2-} + yHCO_3^- + 2H^+ + 2OH^- =$$
$$Ca_{10\text{-}x\text{-}y}[(HPO_4)_x(CO_3)_y(PO_4)_{6\text{-}x\text{-}y}](OH)_{2\text{-}x} + (8\text{-}x)H^+ . \qquad (1)$$

According to the above scheme, HPO_4^{2-} and CO_3^{2-} ions compete for the PO_4^{3-} tetrahedra. CO_3^{2-} ions may also compete for the hydroxyl sites. The charge imbalance created by these substitutions will be compensated by the Ca-vacancies. Monovalent Na^+ and K^+ ions, as well as the divalent Mg^{2+} ions, will also substitute over a certain fraction of the Ca-sites. In the hard tissue nanocrystals Na^+, K^+ and Mg^{2+} ions also participate in the complex chemistry of the formed apatitic CaP (Ap-CaP). Bone mineral, having extremely small sizes of biological apatite crystals (i.e., 20 to 30 nm), contains substantial amounts of CO_3^{2-} (4 to 8 wt%), 0.5% Mg^{2+}, 0.7% Na^+, and is about 10% Ca-deficient with the accompanying increase in reactivity related to this condition [2]. Precipitation of apatitic calcium phosphates may also be accompanied by a slight pH decrease (Eq^n 1). All the precipitated apatitic calcium phosphate powders, therefore, possess, some kind of Ca-deficiency, HPO_4^{2-} and CO_3^{2-} ions in their crystal structures. The addition of 5 mM K^+, 1.5 mM Mg^{2+}, and 0.5 mM SO_4^{2-} ions into a solution as described above may result in the production of either x-ray amorphous CaP (ACP) or cryptocrystalline (*i.e., yielding poor crystallinity XRD patterns incapable of resolving the apatite's quartet of peaks, namely (211), (112), (300) and (202) reflections, over the Cu K_α-radiation 2θ range of 30 to 35°*, some researchers are used to call such a material "poorly crystalline") depending on the synthesis conditions/parameters and depending on how well the pH was maintained constant at 7.4 and 36.5°C. Human blood contains 5 mM K^+, 1.5 mM Mg^{2+} and 0.5 mM SO_4^{2-}. Increasing the amount of Na^+ and Cl^- ions to the levels present in human blood plasma (i.e., 142 and 103 mM, respectively) would just increase the ionic strength of the biomimetic synthesis solution to that of the human blood plasma, which is 149.5 mM. The ionic strength (I) of the human blood plasma is calculated as shown below.

$$I = \frac{1}{2} \left[\overset{Ca^{2+}}{(2.50\times10^{-3})(2)^2} + \overset{Na^+}{(1.42\times10^{-1})(1)^2} + \overset{HPO_4^{2-}}{(1.0\times10^{-3})(2)^2} + \overset{Cl^-}{(1.03\times10^{-1})(1)^2} + \right.$$
$$\left. \overset{HCO_3^-}{(2.7\times10^{-2})(1)^2} + \overset{K^+}{(5\times10^{-3})(1)^2} + \overset{Mg^{2+}}{(1.5\times10^{-3})(2)^2} + \overset{SO_4^{2-}}{(5\times10^{-4})(2)^2} \right] = 0.1495 \text{ M} = 149.5 \text{ mM}$$

Theoretically, if a solution has a low ionic strength, this means that the ionic diffusion will be enhanced in such a solution. In a solution of low ionic strength and high ionic diffusion, more nucleation sites are present for the precipitation reactions. CO_2 is released from an aqueous solution at a faster rate if the solution has a low ionic strength [1]. The presence of NaCl in human blood plasma is for the purpose of adjusting the value of the ionic strength at 149.5 mM. On the other hand, if one increases the ionic strength of a solution to much higher values, such as, 600 mM [3] or 1100 mM [4], its rate of CO_2 release would be slowed down significantly. The presence of 1.5 mM Mg^{2+} in the biomimetic synthesis or biomineralization media is for basically suppressing the growth of well-crystallized apatitic calcium phosphates, favoring instead the formation of cryptocrystalline apatitic calcium phosphates or even ACP in numerous cases [5].

The simplest aqueous solution we developed which can be used in the biomimetic synthesis of hard tissue substitutes or regeneration materials, to replace distilled or deionized water, is the SIEM (saline ionic essentials medium) solution. The composition of SIEM (with a pH value close to that of blood plasma) is given in Table 1, which needs to be prepared in a preboiled deionized water. Heating deionized water to a rolling boil ensures the removal of any dissolved HCO_3^- in it, therefore, it becomes possible to start a careful solution preparation or synthesis procedure by using water free from HCO_3^-. The SIEM solution has an ionic strength of 133 mM, it perfectly matches the Mg^{2+}, K^+, SO_4^{2-}, and HCO_3^- ion concentrations of blood plasma perfectly, has a Na^+ concentration similar to that of blood plasma, and most importantly it is a solution free of any colloidal (i.e., invisible to naked eye) precipitates, it will also not precipitate anything when heated to the physiological temperature of 36.5°C. The solution in

Table 1 is frequently used in our labs, in place of pure (i.e., deionized) water, in synthesizing powders or granules of brushite, amorphous CaP, carbonated Ca-deficient apatite and $CaCO_3$. We also use this solution, instead of pure water, in determining/testing the *in vitro* dissolution rates of various calcium phosphates or polymer/CaP hybrids we synthesize.

Table I   Preparation of 1 liter of SIEM (Saline Ionic Essentials Medium)

Chemical	Amount (g/L)	Ion in SIEM	Concentration (mM)
Na_2SO_4	0.071	Na^+	123
$MgCl_2 \cdot 6H_2O$	0.305	Mg^{2+}	1.5
KCl	0.373	K^+	5
NaCl	5.552	Cl^-	103
$NaHCO_3$	2.268	HCO_3^-	27
		SO_4^{2-}	0.5

If one only replaces, in Table 1, $NaHCO_3$ with 1.84 g of NaOH, then the resulting solution (SIEM-A) will perfectly match the Na^+ (142 mM), Cl^- (103 mM), K^+ (5 mM), Mg^{2+} (1.5 mM) and SO_4^{2-} (0.5 mM) concentrations of the human blood plasma and the SIEM-A solution will have a pH of about 12 at room temperature. The highly novel SIEM-2 solution is especially useful in obtaining *in vitro* (actually in Teflon®) alkaline conditions exactly at the human blood plasma concentrations and we use the SIEM-2 solution to simulate the alkaline phosphatase (ALP) environment in our labs.

The early work of Ringer [6] is perhaps among the most neglected in the science literature, in which he identified the influence of most of the inorganic ions present in blood plasma on the contraction of ventricles. Ringer's original solution possessed 130 mM Na^+, 4 mM K^+, 109 mM Cl^- and 1.4 mM Ca^{2+} at a time when the accurate determination of the ion concentrations of blood plasma was almost impossible [6]. Owing to this neglect, it is sad but not surprising to encounter today some research articles which incorrectly labels the soaking of cotton fibers first in a solution of $(NH_4)_2HPO_4$ and then into a solution of $Ca(NO_3)_2 \cdot 4H_2O$ as "biomimetic synthesis" [7]. Likewise, mixing a solution of chitosan dissolved in acetic acid with a solution of $Ca(NO_3)_2 \cdot 4H_2O$ cannot not be regarded as biomimetic synthesis [8]. Literature is filled with such mislabeled research efforts. Do mammalians have ammonium or nitrate ions in their intra- or extracellular fluids?

BIOMINERALIZATION MEDIA

Table 2 lists the biomineralization media we developed in our labs since 1996 [9]. Each one of these solutions have been shown by us to be effective in different applications, including their use in biomimetic powder, whisker or granule synthesis, in depositing bonelike nanoparticles of calcium phosphates on the surfaces of metals, ceramics and polymers, in transforming the soaked materials (Ti, Ti6Al4V, 316L stainless steel, $CaCO_3$, numerous calcium phosphates, sodium silicate glasses, collagen, cellulose, cross-linked gelatin, cross-linked polyvinyl alcohol hydrogels, etc.) into higher BET surface area end-products or in transforming one calcium phosphate into another at 36.5°C and pH 7.4. The most significant advantage such biomineralization media exhibit is their ability (upon simple soaking runs performed at 36.5°C in sterile glass media bottles) to increase the surface area of the synthetic hard tissue regeneration materials by then allowing enhanced adsorption of blood proteins, blood biomolecules and blood growth factors to the new surfaces. Any synthetic biomaterial, regardless of being a ceramic, metal or polymer, shall indeed pass a very difficult exam within the first 72 h of their *in vivo* implantation. In brief, the surface of any hard tissue regeneration biomaterial which is not suitable for the attachment of blood proteins (lack of any significant surface area (in m^2/g) is one of the main culprits here) and for the osteoblast attachment/proliferation will eventually fail in apposition to the surface of the natural bone tissue.

Table II   Biomineralization media developed [9]

Chemical	Amount (g/L) used to prepare the media in 1 L of water			
	Tris-SBF [10]	*Lac*-SBF [11]	BM-3 [12]	BM-7 [12]
Na_2SO_4	0.071	0.071	-	-
$MgCl_2 \cdot 6H_2O$	0.305	0.305	0.166	0.166
KCl	0.373	0.373	0.398	0.398
NaCl	6.546	5.260	4.787	4.787
$NaHCO_3$	2.268	2.268	3.701	3.701
$CaCl_2 \cdot 2H_2O$	0.368	0.368	0.265	0.333
Na_2HPO_4	0.142	0.142	-	-
$NaH_2PO_4 \cdot H_2O$	-	-	0.125	0.125
Tris	6.057	-	-	-
1 M HCl	35 mL	-	-	-
Na-lactate	-	2.466	-	-
1 M Lactic acid	-	1.5 mL	-	-
Ca/P molar ratio	2.50	2.50	1.99	2.50

The *Tris*-SBF (synthetic body fluid) solution we developed through 1996-1997 [10] is the first and still the only Tris-buffered SBF solution which is capable of strictly matching the HCO_3^- concentration (i.e., 27 mM) of the human blood plasma. The step-by-step preparation recipe of this solution is given elsewhere [9]. Tris-SBF was shown to (*i*) deposit (*via* so called "biomimetic coating") cryptocrystalline apatitic CaP on pure Ti [13] and Ti6Al4V [14-16], (*ii*) transform apatitic CaP, β-TCP ($Ca_3(PO_4)_2$) or biphasic apatitic CaP-TCP whiskers into high surface area (120 m^2/g, BET method) osteoblast proliferating biomaterials [17, 18], and (*iii*) to increase the BET surface area of collagen membranes and sponges by coating those with apatitic, Ca-deficient, carbonated apatite nanoparticles [19] at 36.5-37°C and pH 7.4. *Tris*-SBF (of 27 mM HCO_3^- concentration) is also a versatile aqueous medium to synthesize the powders of x-ray amorphous CaP (ACP) nanoparticles (with mean particle sizes of 60 nm and BET surface area of 250 m^2/g) at 37°C and pH 7.4 [10, 20]. 27 mM HCO_3-Tris-SBF [10] solutions can also be used to *in situ* produce cryptocrystalline, carbonated apatitic CaP nanoparticles upon keeping these solutions in sealed and sterile glass media bottles in a refrigerator (+4°C) for 120 days, as shown in the ATR-FTIR, BET, XRD and TEM data presented in Figure 1. The resulting nanoparticles were found to have amazingly high surface areas (900 m^2/g; Fig. 1). Such high surface area biomaterials (which are composed of apatitic CaP of unquestionable biocompatibility with the human body) are strong candidates for novel drug carrier/delivery applications.

SBF solutions need not always be buffered to the physiological pH (7.4) by using Tris or Hepes. We have developed quite a reliable method of buffering our 27 mM HCO_3-containing SBF solutions by using the urea-enzyme urease pair [21]. In this technique, we first added urea (H_2NCONH_2) to the SBF solution at the concentration of 0.78 M, followed by adding the enzyme urease at different amounts as shown in Figure 2. When one added the enzyme urease to

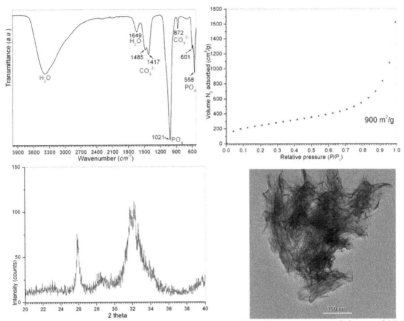

Figure 1.　FTIR, BET surface area, XRD, and TEM data of the *in situ* precipitates of 27 mM HCO₃-Tris-SBF solutions kept in sealed bottles at 4°C for 120 days

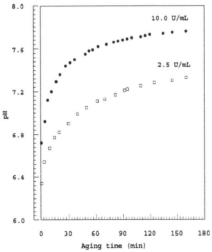

Figure 2.　pH control provided by enzyme urease (shown at two different enzyme concentrations) in urea-containing body fluids at 37°C.

the urea containing SBF solution at the concentration of approximately 7 Units/mL, the enzyme urease then catalyzed the thermal decomposition of urea at 37°C very smoothly and the solution pH was stabilized at 7.4 in less than an hour to remain stable at that pH for the following 1 week of ageing (at 37°C) [21]. To such urea-enzyme urease-containing SBF solutions one can simultaneously add significant amounts of $CaCl_2 \cdot 2H_2O$ and Na_2HPO_4 salts to produce CaP powders at the larger scale. If one adds the same amount of salts to pure SBF, then the pH drops quite rapidly. This is a robust method for the biomimetic synthesis of carbonated nanoparticles of apatitic CaP.

The step-by-step preparation recipe for the *Lac*-SBF solutions is reported elsewhere [9, 11]. *Lac*-SBF solutions do not use Tris or Hepes, but they employ the pair of Na-lactate and 1 M lactic acid (only 1.5 mL of lactic acid is used in preparing 1 L of *Lac*-SBF, this is a very small amount) to adjust the solution pH at the physiological value of 7.4 and at 37°C. The *Lac*-SBF solution was found to be capable of producing a biomimetic coating on the alkali-treated (5 M NaOH solution, 60°C, 24 h soaking) pure Ti substrates as shown in Figure 3. The undercoat seen

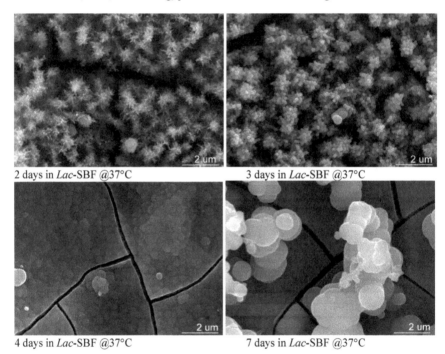

2 days in *Lac*-SBF @37°C   3 days in *Lac*-SBF @37°C

4 days in *Lac*-SBF @37°C   7 days in *Lac*-SBF @37°C

Figure 3.   Progress of biomimetic coating on the surfaces of alkali-treated Ti (not Ti6Al4V) substrates (10 x 10 x 1 mm) soaked in 75 mL of *Lac*-SBF solutions at 37°C.

in 2 and 3 days samples are noticeably different than those produced by 27 mM HCO_3-Tris-SBF solutions [14-16]. However, the samples kept in solution for 4 and 7 days were quite similar in morphology to the 27 mM HCO_3-Tris-SBF coatings. Based on the evidence provided in the data of Fig.3, Lac-SBF solutions can be used in place of Tris-buffered SBF solutions. Some opinions against the use of Tris-buffered SBF solutions in testing the bioactivity of synthetic biomaterials are already in place. For instance, a recent article by Boccaccini *et al.* [22] disclosed that the Tris-buffer present in the SBF (simulated/synthetic body fluid) solutions was causing an increased dissolution of the surface constituents of soaked bioglass and glass-ceramics samples and therefore led to the premature crystallization of apatite on sample surfaces, tarnishing the reliability of those so-called *in vitro* bioactivity measurements based on such SBF solutions. Tris is known to be a calcium chelator [23].

Lac-SBF solutions can also be used in the biomimetic transformation (at 37°C) of brushite ($CaHPO_4 \cdot 2H_2O$) into biphasic octacalcium phosphate (OCP)-Ca-deficient apatitic CaP, as shown in Fig. 4. Brushite is a mildly acidic calcium phosphate compound, although it can transform rapidly into the bone mineral when implanted [24]. The biomimetic transformation of brushite (by soaking in a *Lac*-SBF solution) into biphasic OCP-HA before its implantation may circumvent any tissue inflammation problems associated with the direct use of brushite powders, cements or granules [25]. The change in morphology of brushite particles soaked in *Lac*-SBF solutions at 37°C is shown in the SEM images of Fig. 5. The BET surface area of brushite powders increased from about 1 m²/g (as is powders) to 120 m²/g upon soaking for 1 week in the Lac-SBF solutions. This increase in surface area is one of the key advantages of such biomimetic transformations performed in a solution such as Lac-SBF (which perfectly matches all the inorganic ion concentrations of the human blood plasma). Brushite is a very easy-to-synthesize CaP even in industrial large-scale proportions, and this study showed a robust way of turning such inexpensive bioactive and non-cytotoxic brushite into a biomaterial with a significantly large surface area. Large surface area biomaterials can be used in drug delivery applications especially in cancer-related research. The particle sizes of these new biomaterials are also large enough not to migrate through the blood-brain barrier (BBB).

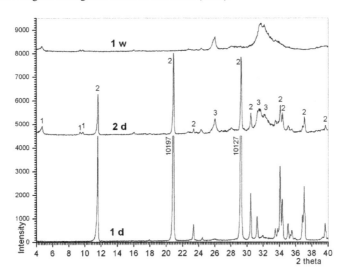

Figure 4.   XRD data of 1 g brushite powders soaked in 75 mL of *Lac*-SBF solutions at 37°C for different periods (1 day to 1 week), 1: OCP, 2: brushite and 3: HA peaks.

2 days   7 days

Figure 5.   The SEM images of 1 g of brushite powder soaked in 75 mL Lac-SBF solution for 2 days and 1 week at 37°C.

BM-3 and BM-7 solutions of Table II are biomineralization media we developed based on the inorganic ion concentrations of DMEM (Dulbecco's Modified Eagle Medium) solutions which are extensively used as media in performing the cell culture studies. BM-3 solution has a Ca/P molar ratio of 1.99 just like the DMEM solutions. It shall be remembered that DMEM solutions also contain amino acids, vitamins, glucose and Hepes, our solutions, on the other hand, do not contain any of these, therefore, these are very easy-to-prepare in large volumes. BM-7 solution has a Ca/P molar ratio of 2.5 similar to the human blood plasma. BM-3 solutions of pH 7.4 were found to have the unique ability of transforming brushite soaked in those (no stirring or agitation needed) at 37°C to single-phase octacalcium phosphate (OCP) in less than 72 hours [12], as shown in Fig. 6. BM-3 solutions thus provide a simple alternative to the biomimetic synthesis (i.e., in a solution containing Na^+, K^+, Ca^{2+}, Mg^{2+}, HCO_3^-, HPO_4^{2-} and Cl^- ions just like the human blood plasma) of OCP. Previous literature is not able to show a synthesis method for OCP similar to this one. BM-7 solutions were found to have the similar ability of transforming brushite into octacalcium phosphate, but the crystallinity of OCP particles obtained in BM-7 solutions were much less in comparison to those obtained in BM-3 solutions.

Figure 6.   OCP ($Ca_8(HPO_4)_2(PO_4)_4 \cdot 5H_2O$) crystals synthesized at 37°C in less than 72 h in 75 mL of BM-3 solution by starting with 1 g of brushite powders.

BM-7 solutions were found to coat x-ray amorphous CaP (ACP) on glass and pure Ti substrates. Figure 7a shows two glass slides; one of them was immersed into the BM-7 solution for only 24 h at 37°C. Figure 7b shows the SEM image of the ACP-coated bottom slide shown in Fig. 7a. To the best of our knowledge, there have been no solution recipes available in the previous literature with the ability of coating ACP. Figure 8 showed the SEM, XRD, and FTIR data for the ACP deposits obtained on alkali-treated Ti soaked in BM-7 solution for 24 h, 37°C.

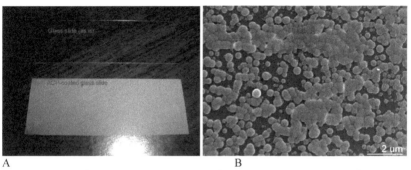

A   B

Figure 7.   Digital camera (7a) and SEM (7b) images of ordinary glass slides coated with amorphous CaP (ACP) upon soaking in the BM-7 solution for 24 h at 37°C.

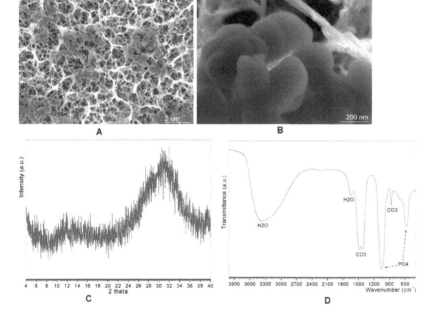

Figure 8.   SEM (a) and (b), XRD (c), and FTIR (d) data for alkali-treated Ti soaked in BM-7 solution for 24 h at 37°C.

Soaking of brushite granules in BM-7 solutions (72 h, 37°C) increased their BET surface area from around 1 to 115 m^2/g [11]. The SEM images of Fig. 9 depicted the nanotextured surface formed after BM-7 soaking. Such examples of surface engineering on brushite particles or granules were not reported before.

Figure 9.   SEM images depicting the change in surface texture of brushite soaked in BM-7 solutions for 72 h at 37°C ((a) *as is*, (b) and (c) upon soaking in BM-7 for 72 h).

The major advantage with the BM-3 and BM-7 solutions [12], over those of 27 mM HCO_3-*Tris*-SBF and *Lac*-SBF (Table II), is obviously the simplicity of the solution preparation. They require no Tris, Hepes or Na-lactate buffering. One adds all the chemicals, one by one, into deionized water and the solution instantly becomes ready to use. On the other hand, SBF solutions do require careful pH adjustments by using either Tris-HCl or Na-lactate-lactic acid pairs. Moreover, in preparing SBF solutions one first encounters a solution turbidity (caused by the formation of colloidal calcium phosphate precipitates) and that turbid solution then turned into a transparent one either by slow and careful additions of 1 M HCl (35 to 40 mL in case of SBF) or 1 M lactic acid (1.5 mL in case of *Lac*-SBF) solutions. During the SBF preparations, the operator must stop adding the acid right at the moment where the pH-meter reads 7.4. This also places a burden on the accurate calibration of the pH meter just prior to the SBF preparation. If the person (typically a less-experienced student or a young researcher lacking the solution preparation skills) preparing the SBF solution stops adding the acid a bit late, e.g., stopping when the pH was already dropped to 7.25, and if that person still wishes to use this solution as the SBF, his/her results would be significantly influenced or altered by this low pH of the solution. BM-3 and BM-7 solutions, therefore, do not present any such difficulties as long as the person in charge of preparing such solutions has a carefully calibrated analytical balance available to weigh the chemicals of Table II.

It is also possible to synthesize x-ray amorphous calcium phosphate (ACP) powders in solutions which (i) mimic the ion concentrations of the human blood plasma and (ii) are free of any buffering agents such as Tris or Hepes. This is accomplished by using Ca metal (elemental) as the only Ca source during syntheses [9, 26]. ACP powders are biomimetically synthesized as follows; 0.187 g KCl, 0.153 g $MgCl_2 \cdot 6H_2O$, 2.776 g NaCl, 1.134 g $NaHCO_3$ and 0.355 g Na_2HPO_4 are dissolved, one by one, in 500 mL of deionized water, and then 0.251 g Ca metal is added to the above solution. The solution needs to be stirred at room temperature for 30 min. The formed precipitates are removed from the solution by filtering, washed with water and dried at room temperature. ACP powders consist of 50 nm diameter spherical particles, as shown in Fig. 10.

Figure 10.   (a) Biomimetic ACP nanopowders synthesized by using Ca metal

Figure 10. (b) Biomimetic ACP nanopowders synthesized by using Ca metal

Similarly, cryptocrystalline apatitic CaP powders are biomimetically synthesized as follows; 0.187 g KCl, 0.153 g MgCl$_2$·6H$_2$O, 2.776 g NaCl, 1.134 g NaHCO$_3$ and 0.71 g Na$_2$HPO$_4$ are dissolved, one by one, in 500 mL of deionized water, and then 1.838 g CaCl$_2$·2H$_2$O is added to the above solution [9, 26]. The solution needs to be stirred at room temperature for 30 min. The formed precipitates are removed from the solution by filtering, washed with water and dried at room temperature. Cryptocrystalline powders consist of 150 nm diameter spherical particles, as shown in Fig. 11.

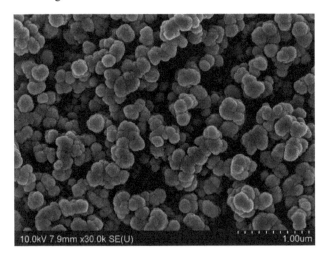

Figure 11. (a) Biomimetic cryptocrystalline CaP nanopowders synthesized by adding CaCl$_2$·2H$_2$O to a SIEM solution

Figure 11.   (b) Biomimetic cryptocrystalline CaP nanopowders synthesized by adding $CaCl_2 \cdot 2H_2O$ to a SIEM solution

CONCLUSION

Five separate biomineralization media (SIEM, 27 mM HCO_3-*Tris*-SBF, *Lac*-SBF, BM-3, and BM-7) are reported. All of these media can be used, in place of distilled or deionized water, in synthesizing biomimetic ceramics, polymers or hybrids of ceramics-polymers (ceramers). All of these media can also be used in determining the solubility or dissolution rates of newly-synthesized biomaterials, since these solutions try to mimic the inorganic ion composition the human blood plasma. Most of these media can be used in increasing the BET surface area of newly synthesized biomaterials. The examples provided here showed that metal or ceramic surfaces can be altered upon soaking in such media. Although the conventional SBF recipe was mainly promoted over the last two decades to test the so-called *in vitro* bioactivity of synthetic biomaterials, biomineralization media and the more advanced SBF solutions reported here shall be regarded as versatile tools for surface engineering of biomaterials for numerous clinical applications. We do not suggest the use of SBF solutions in testing the *in vitro* bioactivity of biomaterials since the best medium to test the *in vitro* bioactivity of biomaterials is the "cell-free cell culture solutions," such as DMEM or α-MEM (minimum essentials medium), which are already available as sterile solutions from different vendors all over the world.

REFERENCES
[1] P. A. A. P. Marques, M. C. F. Magalhaes, and R. N. Correia, Inorganic Plasma with Physiological CO_2/HCO_3^- Buffer, *Biomaterials*, **24**, 1541-48 (2003).
[2] A. S. Posner and F. Betts, Synthetic Amorphous Calcium Phosphate and Its Relation to Bone Mineral Structure, *Acc. Chem. Res.*, **8**, 273- (1975).
[3] A. C. Tas, Electroless Deposition of Brushite ($CaHPO_4 \cdot 2H_2O$) Crystals on Ti-6Al-4V at Room Temperature, *Int. J. Mater. Res.*, **97**, 639-44 (2006).
[4] A. C. Tas and S. B. Bhaduri, Rapid Coating of Ti6Al4V at Room Temperature with a Calcium Phosphate Solution similar to 10X Simulated Body Fluid, *J. Mater. Res.*, **19**, 2742-49 (2004).
[5] P. Habibovic, F. Barrere, C. A. van Blitterswijk, K. de Groot, and P. Layrolle, Biomimetic Hydroxyapatite Coating on Metal Implants, *J. Am. Ceram. Soc.*, **85**, 517-22 (2002).
[6] S. Ringer, Concerning the Influence Exerted by Each of the Constituents of the Blood on the Contraction of the Ventricle, *J. Physiol.*, **3**, 380-93 (1882).

[7] C. Xing, S. Ge, B. Huang, Y. Bo, D. Zhang, and Z. Zheng, Biomimetic Synthesis of Hierarchical Crystalline Hydroxyapatite Fibers in Large-scale, *Mater. Res. Bull.*, **47**, 1572-6 (2012).

[8] J. Li, H. Sun, D. Sun, Y. Yao, F. Yao, and K. Yao, Biomimetic Multicomponent Polysaccharide/Nano-hydroxyapatite Composites for Bone Tissue Engineering, *Carbohyd. Polym.*, **85**, 885-94 (2011).

[9] http://www.cuneyttas.com

[10] A. C. Tas, Synthesis of Biomimetic Ca-Hydroxyapatite Powders at 37°C in Synthetic Body Fluids, *Biomaterials*, **21**, 1429-38 (2000).

[11] M. A. Miller, M. R. Kendall, M. K. Jain, P. R. Larson, A. S. Madden, and A. C. Tas, Testing of Brushite ($CaHPO_4 \cdot 2H_2O$) in Synthetic Biomineralization Solutions and *In Situ* Crystallization of Brushite Micro-Granules, *J. Am. Ceram. Soc.*, **95**, 2178-88 (2012).

[12] N. Temizel, G. Girisken, and A. C. Tas, Accelerated Transformation of Brushite to Octacalcium Phosphate in New Biomineralization Media between 36.5° and 80°C, *Mater. Sci. Eng. C*, **31**, 1136-43 (2011).

[13] C. Kim, M. R. Kendall, M. A. Miller, C. L. Long, P. R. Larson, M. B. Humphrey, A. S. Madden, and A. C. Tas, Comparison of Titanium soaked in 5 M NaOH or 5 M KOH Solutions, *Mater. Sci. Eng. C*, **33**, 327-39 (2013).

[14] S. Jalota, S. B. Bhaduri, and A. C. Tas, Effect of Carbonate Content and Buffer Type on Calcium Phosphate Formation in SBF Solutions, *J. Mater. Sci. Mater. M.*, **17**, 697-707 (2006).

[15] S. Jalota, S. B. Bhaduri, and A. C. Tas, Osteoblast Proliferation on Neat and Apatite-like Calcium Phosphate-coated Titanium Foam Scaffolds, *Mater. Sci. Eng. C*, **27**, 432-40 (2007).

[16] S. Jalota, S. Bhaduri, S. B. Bhaduri, and A. C. Tas, A Protocol to Develop Crack-Free Biomimetic Coatings on Ti6Al4V Substrates, *J. Mater. Res.*, **22**, 1593-1600 (2007).

[17] S. Jalota, S. B. Bhaduri, and A. C. Tas, In vitro testing of Calcium Phosphate (HA, TCP, and biphasic HA-TCP) Whiskers, *J. Biomed. Mater. Res.*, **78A**, 481-90 (2006).

[18] A. C. Tas, Formation of Calcium Phosphate Whiskers in Hydrogen Peroxide (H_2O_2) at 90°C, *J. Am. Ceram. Soc.*, **90**, 2358-62 (2007).

[19] S. Jalota, S. B. Bhaduri, and A. C. Tas, Using a Synthetic Body Fluid (SBF) Solution of 27 mM HCO_3^- to Make Bone Substitutes More Osteointegrative, *Mater. Sci. Eng. C*, **28**, 129-40 (2008).

[20] D. Bayraktar and A. C. Tas, Chemical Preparation of Carbonated Calcium Hydroxyapatite Powders at 37°C in Urea-containing Synthetic Body Fluids, *J. Eur. Ceram. Soc.*, **19**, 2573-9 (1999).

[21] D. Bayraktar and A. C. Tas, Formation of Hydroxyapatite Precursors at 37°C in Urea- and Enzyme Urease-containing Body Fluids, *J. Mater. Sci. Lett.*, **20**, 401-3 (2001).

[22] D. Rohanova, A. R. Boccaccini, D. M. Yunos, D. Horkavcova, I. Brezovska, and A. Helebrant, Tris Buffer in Simulated Body Fluid Distorts the Assessment of Glass-Ceramic Scaffold Bioactivity, *Acta Biomater.*, 7, 2623-30 (2011).

[23] P. Premoli, A. Manzoni, and C. Calzi, Method for the Quantitative Determination of Calcium in Biological Fluids through Direct Potentiometry, European Patent No: EP 0153783 B1, October 30, 1991.

[24] M. Kuemmerle, A. Oberle, C. Oechslin, M. Bohner, C. Frei, I. Boecken, and B. von Rechenberg, Assessment of the Suitability of a New Brushite Calcium Phosphate Cement for Cranioplasty – An Experimental Study in Sheep, *J. Craniomaxillofac. Surg.*, **33**, 37–44 (2005).

[25] A. C. Tas, Granules of Brushite and Octacalcium Phosphate from Marble, *J. Am. Ceram. Soc.*, **94**, 3722-6 (2011).

[26] A. C. Tas, Calcium Metal to Synthesize Amorphous or Cryptocrystalline Calcium Phosphates, *Mater. Sci. Eng. C*, **32**, 1097-1106 (2012).

STRUCTURAL CHARACTERIZATION AND COMPRESSIVE BEHAVIOR OF THE BOXFISH HORN

Wen Yang[1*], Vanessa Nguyen[2], Michael M. Porter[1], Marc A. Meyers[1,2,3], Joanna McKittrick[1,2]

[1]Materials Science and Engineering Program, [2]Department of Mechanical and Aerospace Engineering, [3]Department of Nanoengineering
University of California, San Diego, La Jolla, CA 92093, USA
[*] Corresponding author, email: wey005@eng.ucsd.edu

ABSTRACT
    Boxfish have a rigid carapace that restricts body movement making them slow swimmers. Some species of boxfish (*Lactoria cornuta*) have lightweight horns that function as a form of defense. The boxfish horns are nearly hollow and have an intricate hierarchical structure. The structural organization and compressive properties of the boxfish horns are described here to understand the mechanical behavior and damage mechanisms.

INTRODUCTION
    Boxfish, belonging to the family *Ostraciidae*, are named after their unique boxy profile, which is generally an oblong, triangular or square-like shape. There are over twenty species of fish in the family *Ostraciidae*, which live in the Atlantic, Pacific and Indian Oceans [1,2]. Found in warm tropical and subtropical waters, they typically inhabit coral reefs and grass beds at depths of 1 to 45 m [3]. They are usually yellow; however, their color can vary, ranging from orange, green, blue or white [4]. They also secrete toxins into the surrounding water to deter predators [5]. Their body, growing up to 50 cm in length, is composed of a rigid carapace that restricts significant body movement [6-8]. The carapace is composed of several highly mineralized [9], hexagonal dermal scutes (bony plates), with some incidence of heptagons and pentagons (shown in Figure 1a). As a result, they are relatively slow swimmers that glide or hover through the water - a method of locomotion known as ostraciiform swimming [10]. The boxy shape and rigid carapace allow the boxfish to swim in this manner by minimizing vortices and drag [8].
    One species, *Lactoria cornuta*, is called the longhorn cowfish because of the two long horns that protrude from the head. Another pair of horns are located below the caudal fin and protrude from the posterior. Khan et al. [11] suggested that the horns are a type of defense mechanism, providing obstructions that discourage other fish from swallowing them. They also state that the horns can be regenerated in a few months if broken off, indicating the horns are a necessary adaptation developed for protection from predators.
    The boxfish scutes and fish scales are composed of hydroxyapatite ($Ca_{10}(PO_4)_6(OH)_2$) and type I collagen [9, 12, 13]. Similarly, the horns should also be composed of mineralized collagen fibers. The arrangement of the mineralized fibers significantly affects the mechanical properties in bony tissues [14].
    The purpose of this work is to analyze the mechanical properties and structure of the boxfish horns. Understanding the mechanical behavior and damage mechanisms of these horns may provide inspiration for lightweight synthetic materials for structural or defense applications.

EXPERIMENTAL TECHNIQUES
    Two boxfish (deceased), *Lactoria cornuta*, were obtained from Scripps Institution of Oceanography at University of California, San Diego and preserved in 70% isopropanol. The

structure of the boxfish horns was examined by 3D digital optical microscopy, scanning electron microscopy (SEM), and micro-computed tomography (μ-CT). The compressive mechanical properties and fracture mechanisms were investigated by compressing a small section of a single horn.

A whole boxfish, including its horns, was scanned with a Skyscan 1076 (Kontich, Belgium) μ-CT scanner. For sample preparation, the boxfish was wrapped in tissue paper moistened with a phosphate buffer saline solution and placed in a sealed tube to prevent the specimen from drying out during scanning. An isotropic voxel size of 36 μm, an electric potential of 100kV, and a current of 100 μA was applied during scanning using a 0.5 mm aluminum filter with a rotation step of 0.6 degrees and exposure time of 90 ms. A beam hardening correction algorithm was applied during image reconstruction. Images and 3D rendered models were analyzed using Skyscan's Data Viewer and CTVox software.

To characterize the structure, two sections of the horn were cut - one along the longitudinal direction and one in the transverse (cross-sectional) direction. Both sections were partially deproteinized in a 6% NaClO solution for 15 min to better visualize the microstructures [15].The partially deproteinized horn sections were observed with a digital optical microscope (VHX-1000, Keyence, NJ) and environmental scanning electron microscope (FEI-XL30, FEI Company, Hillsboro, OR). Before SEM observation, the horn sections were fixed in a 2.5% glutaraldehyde solution for 3 hrs, then immersed in a gradient ethanol series (30 vol.%, 50 vol.%, 75 vol.%, 80 vol.%, 95 vol.% to 100 vol.%) to remove water from the specimens while preventing shrinkage due to dehydration. The samples immersed in ethanol were dried in a critical point dryer (Auto Samdri 815A, Tousimis, MD) to preserve the original shape of the horns. The dried samples were then sputter coated with iridium using an Emitech K575X sputter coater (Quorum Technologies Ltd., West Sussex, UK) and examined by SEM.

A small section (1.5 mm in length) of a horn was cut and loaded in compression using a load frame (Instron 3342, Norwood, MA). The diameter of the sample gradually decreased from the root to the tip, from ~1.75 mm to ~1.6 mm. As the sample was short, no attachments were fixed on the sample. The sample was immediately tested after removal from the isopropanol and loaded longitudinally with a strain rate of 10^{-3} sec^{-1}.

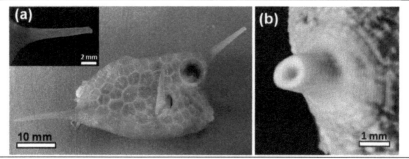

Figure 1. (a) Photograph of the dextral view of a boxfish (*Lactoria cornuta*) and one horn (top left); (b) micro-computed tomography image showing the transverse cross-section of a horn.

STRUCTURAL CHARACTERIZATION OF THE BOXFISH HORN

Figure 1a shows the dextral side of the boxfish. The length of the whole boxfish, including the horns and tail, is ~60 mm. The horns of the fish are ~12 mm long and appear hollow in the μ-CT image (Figure 1b), which demonstrates that the horn sheath is mineralized.

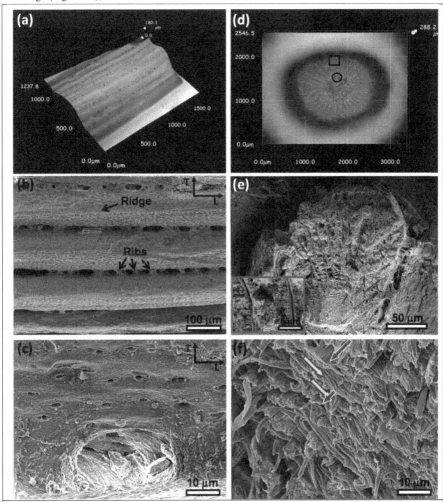

Figure 2. Images of the outer and inner surfaces of the horn. (a) 3D digital microscopic image of the partially deproteinized outer surface showing a petal-like feature composed of ridges; (b) SEM image showing ribs connecting the ridges (L and T in the coordinate system represents the longitudinal and transverse directions, respectively); (c) SEM image of small pores on the outer surface; (d) 3D digital microscopic image of the transverse cross-section; (e) SEM images of fibers extending through the petal-like feature similar to the veins of a leaf from the square region in (d); (f) SEM image of the cross-fiber pattern from circle region in (d).

Figure 2 shows the structural hierarchy of a selected region in the middle of a partially deproteinized horn. The outer surface is composed of several arch-like ridges oriented in the longitudinal direction (Figure 2a).The average diameter of each ridge is ~100 μm (Figure 2b). Between the longitudinal ridges are perpendicularly oriented fibers (ribs) that connect adjacent ridges with a spacing of 10-100 μm (Figure 2b). The ridge surfaces contain several small pores ~3 μm in diameter (Figure 2c). Figure 2d shows the horn cross-section. The outer and inner diameters are 1.8 and 0.8 mm, respectively. The edges of each ridge (see Figures 2a and 2b) extend from the outer surface towards the center, creating a ribbed structure that resembles the petals of a flower. The central core is filled with a low-density organic network near the base (Figure 2d). Figure 2e shows higher magnification of the outer edge of a single rib (square in Figure 2d). The fibers extend through the structure from the center to the edge with ~10 μm spacing between fiber bundles (left bottom image in Figure 2e). Figure 2f shows higher magnification of the inner edge of a single rib (circle in Figure 2d), which has two orientations of fibers that are perpendicular to each other. These hierarchical structures appear to make the horns lightweight and rigid.

The density of the central matrix of the horn decreases from the base to the tip. Figure 3a shows across-section of the horn ~ 3mm from the tip. It is clear that the central region is nearly hollow. The ribbed units contain 2-5 tubules aligned in the longitudinal direction. The fibers appear to be aligned circumferentially around the pores/tubules, as shown in Figure 3b. The average diameter of the tubules changes from 44 μm (close to core) to 18 μm (close to periphery). This gradation in pore diameter has been observed in the foam structure of porcupine quill [16] and channels in the structure of sucker rings from *Dosidicus gigas* [17]. Although the horn is mineralized, due to the extensive porosity, the overall density (by weight and dimension measurements) is ~ 1000 kg/m^3. Nature creates larger tubules/channels/cells in the center to reduce the density but maintain bending resistance. The center of a cylinder does not experience substantial bending stress, thus the material can be removed without compromising the bending resistance. Although bending tests would enhance the analytical results, in this work, due to the limited number of horns and the small sample size, these tests were not performed. However, this unidirectional alignment indicates that the boxfish horn should be stronger when loaded in the longitudinal direction than in other directions [13].

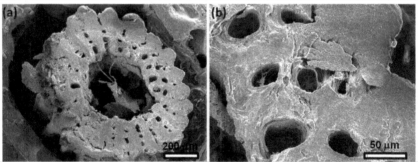

Figure 3. (a) Cross-section of horn close to the tip, (b) several tubules running longitudinally through the horn.

COMPRESSIVE BEHAVIOR AND DAMAGE MECHANISMS

Since the boxfish horn is used for protection and defense, it may be subjected to compressive loading. Therefore, a small section of the horn was selected to investigate the compressive behavior and evaluate the damage mechanisms. An image of this section is shown in Figure 4. The stress-strain curve (Figure 4) was calculated assuming the horn is a hollow cylinder with a wall thickness of ~500 μm. For this sample, the stiffness of the horn is ~430 MPa and the failure strength is ~80 MPa at a strain of ~30%. The horn demonstrated high toughness, as evidence by the considerable plastic deformation experienced before failure, which may involve multiple deformation modes.

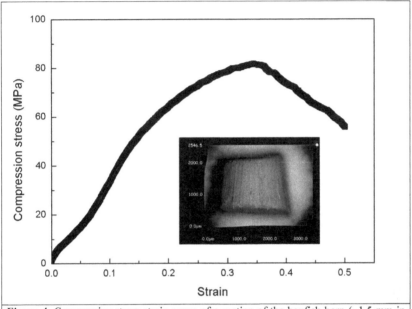

Figure 4. Compressive stress-strain curve of a section of the boxfish horn (~1.5 mm in length). The inset shows an image of the sample before compression.

After compression testing, the fracture surfaces were examined by SEM. The sample fractured between the ridges, as shown in Figure 5a. Some of the ridges buckled (arrow in Figure 5a). At the fractured edges, large fiber bundles pulled out, revealing more clearly the thin fibers connecting the bundles to the matrix (arrows in Figure 5b). Before fracture, most of the fibers showed considerable plastic deformation in the form of necking (arrows in Figure 5c). The fiber bundles also underwent delamination (square in Figure 5d) and buckling (arrow in Figure 5d). As the spacing between ridges deformed, the network of fibers connecting them (ribs) remained mostly intact. All these mechanisms contribute to high energy absorption under compressive loading, providing the horn the necessary toughness to function as a protective appendage.

In summary, the density of the pores/tubules in the boxfish horn decreases from the core to the periphery. As shown in Figure 3, it is believed that the fibers are aligned circumferentially around the pores in the structure. The alignment of the tubules most likely makes the boxfish horn stronger in the longitudinal directions compared to the other orthogonal directions[13]. Compared to the compressive strength of bovine cortical bone (150~180 MPa) the boxfish horn is about half (~ 80 MPa) in the longitudinal direction. However, the density of boxfish horn (1000 kg/m^3) is lower than that of bovine cortical bone (2100 kg/m^3)[19], therefore the strength per unit weight of the boxfish horn is almost the same as cortical bone. The high strength per unit weight results from other structure features. The horn has ribs in the circumferential direction that connect the ridges, and under loading the ribs remain intact to retain structural integrity. Thus, the architecture of the boxfish horn is a model for biomimetic/bioinspired materials design, as it possesses low density, high strength and high toughness.

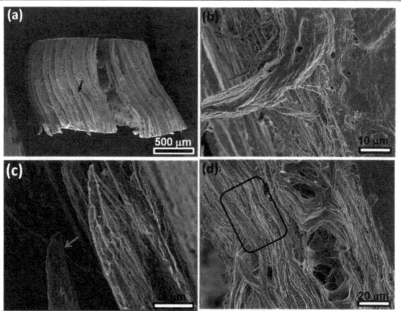

Figure 5. SEM images of (a) horn section after compression showing the primary crack formed between the ridges; (b) fracture surface showing fiber pull-out and stretching; (c) fibers after fracture illustrating fiber necking; (d) fracture surface showing fiber delamination (square) and buckling (arrow).

CONCLUSIONS

This study reports on the structural characterization, compression behavior, and damage mechanisms of the boxfish (*Lactoria cornuta*) horn, which are reviewed as follows:

1. The horn is composed of a mineralized conical sheath that has an organic network inclusion. The density of the organic inclusion decreases from the base to the tip of the horn, becoming nearly hollow.

2. The sheath, which provides most of the horn support, has a hierarchical structure. It has several petal-shaped units on the outer surface that are composed of ridges. The ridges are connected by ribs oriented roughly perpendicular to the ridges. Each ribbed unit is

composed of fibers that extend from the center core to the outer surface. Small tubules (10-50 μm) run longitudinally through the outer sheath.
3. The compressive strength of the horn is ~80 MPa at a failure strain of ~30%.
4. Under compressive loading, longitudinally oriented fibers delaminate, buckle, and neck before fracture, demonstrating that the horn has an energy dissipative structure.
5. The horn has a similar compressive strength to weight ratio as bovine cortical bone.

ACKNOWLEDGEMENTS
We thank Prof. Phil Hastings and H.J. Walker at the Scripps Institute of Oceanography for providing the boxfish. Prof. Robert Sah and Esther Cory are thanked for performing and helping interpret the micro-computed tomography. This work is supported by the National Science Foundation, Ceramics Program Grant 1006931 and UC Lab Research Program, 09-LR-06-118456-MEYM.

REFERENCES
1. F. Santini, L. Sorenson, T. Marcroft, A. Dornburg and M.E. Alfaro, "A multilocus molecular phylogeny of boxfishes (Aracanidae, Ostraciidae; Tetraodontiformes),"*Mol. Phylogenet.Evol.*,**66**, 153-160 (2013).
2. M.A. Ambak, M.M. Isa, M.Z. Zakaria, and M.B. Ghaffar, Fishes of Malaysia, 1st Ed., Terengganu: University Malaysia Terengganu Press (2010).
3. J.B. Hutchins, Ostraciidae. In FAO species identification sheets for fishery purposes. Western Indian Ocean (Fishing area 51), Vol. 3, edited by W. Fischer and G. Bianchi, 9. Rome: FAO (1984).
4. http://marinebio.org/species.asp?id=1463 Longhorn Cowfishes, *Lactoria cornuta*, 14 January 2013.
5. K. Tachibana, "Chemical defense in fish," in Bioorganic Marine Chemistry, P.J. Scheuer, Ed., **2**, 117-138 (1988).
6. J.R. Hove, L.M. O'Bryan, M.S. Gordon, P.W. Webb and D. Weihs, "Boxfishes (Teleostei: Ostraciidae) as a model system for fishes swimming with many fins: kinematics," *J. Exp. Biol.*, **204**, 1459–1471(2001).
7. M.S. Gordon, J.R. Hove, P.W. Webb, and D. Weihs, "Boxfishes as unusually well controlled autonomous underwater vehicles," *Physiol. Biochem.Zool.*,**73**, 663–671 (2001).
8. I.K. Bartol, M. Gharib, P.W. Webb, D. Weihs, and M.S. Gordon, "Body-induced vortical flows: a common mechanism for self-corrective trimming control in boxfishes," *J. Exp. Biol.*, **208**, 327–344 (2005).
9. L. Besseau and Y. Bouligand, "The twisted collagen network of the box-fish scutes,"*Tiss.& Cell*, **30**, 251–260 (1998).
10. W. Blake, "Ostraciiform locomotion,"*J. Mar. Biol. Assoc. UK*, **57**, 1047-1055 (1977).
11. M.S.K Khan, MAM Siddique, MA Haque, "New record of the longhorn cowfish *Lactoria cornuta* (Linnaeus 1758) from inshore waters of the Bay of Bengal, Bangladesh," *Zool. and Ecol.*, **23**, 88-90 (2013).
12. T. Ikoma, H. Kobyahi, J. Tanaka, D. Walsh, S. Mann, "Microstructure, mechanical, and biomimetic properties of fish scales from *Pagrus major*,"*J. Struct. Biol.*,**142**, 327-333 (2003).
13. W. Yang, B. Gludovatz, E.A. Zimmermann, H.A. Bale, R.O. Ritchie and M.A. Meyers, "Structure and fracture resistance of alligator gar (*Atractosteus spatula*) armored fish scales,"*Acta Biomater.*,**9**, 5876-5889 (2013).

14. S. Weiner, W. Traub, H.D. Wagner, "Lamellar bone: structure-function relationships,"*J. Struct. Biol.*, **126**, 241-255 (1999).

15. A.B. Castro-Ceseña, E. Novitskaya, P.-Y. Chen, M. del Pilar Sánchez-Saavedra, G. Hirata, and J. McKittrick, "Comparison of demineralized and deproteinized bone,"*Mater. Res. Soc. Symp. Proc.*, **1301**, 2011, doi: 10.1557/opl.2011.194.

16. W. Yang, J. McKittrick, "Separating the influence of the cortex and foam on the mechanical properties of porcupine quills," *Acta Biomater.*, 2013 (in press), doi: 10.1016/j.actbio.2013.07.004

17. A. Miserez, J.C. Weaver, P.B. Pedersen, T. Schneeberk, R.T. Hanlon, D. Kisailus, H. Birkedal, "Microstructural and biochemical characterization of the nanoporous sucker rings from *Dosidicus gigas*," *Adv. Mater.*,**21**, 401-406 (2009).

18. Z. Manilay, E. Novitskaya, E. Sadovnikov, J. McKittrick, "A comparative study of mature and young cortical bone," *Acta Biomater.*, **9**, 5280-5288 (2013).

19. J.D. Currey, "Mechanical properties of bone tissues with greatly differing functions," *J. Biomech.*,**12**, 313-319 (1979).

COMPARATIVE EVALUATION of CRYSTALLIZATION BEHAVIOR, MICRO STRUCTURE PROPERTIES and BIOCOMPATIBILITY of FLUORAPATITE-MULLITE GLASS-CERAMICS

S. Mollazadeh[1], A. Youssefi[2], B. Eftekhari Yekta[1], J. Javadpour[1], T.S. Jafarzadeh[3], M. Mehrju[4] and M.A. Shokrgozar[4]
[1] School of Metallurgy and Materials Engineering, Iran University of Science & Technology, Tehran, Iran
[2] Par-e-Tavous Research Center, Mashhad, Iran
[3] School of Dentistry, Tehran University of Medical Science, Tehran, Iran
[4] National Cell Bank of Iran- Pasteur Institute of Iran

ABSTRACT

The growing trend for restorative glass-ceramic materials has pushed on to the development of the novel dental glass-ceramic systems. Improved biocompatibility, adequate strength, chemical and wear resistance and excellent aesthetic are the main criteria that make these materials clinically successful. The aim of the present study was to investigate the effect of small additions of TiO_2, ZrO_2, BaO and extra amounts of silica on the microstructural changes and biological properties of an apatite- mullite base glass-ceramic system. Glass transition temperatures and crystallization behavior were investigated using differential thermal analysis (DTA). Addition of TiO_2, ZrO_2, BaO and extra amounts of silica to the base glass led to some changes in the crystallization temperatures and morphology of the crystalline phases. DTA results showed that while TiO_2 and BaO were effective in decreasing the crystallization temperature of the fluorapatite and mullite crystalline phases, ZrO_2 and the extra amounts of SiO_2 increased the crystallization temperature. X-ray diffractometry (XRD) and scanning electron microscopy (SEM) revealed that the precipitated crystalline phases were fluorapatite [$Ca_{10}(PO_4)_6F_2$] and mullite [$Al_6Si_2O_{13}$], which apart from the extra bearing SiO_2 specimen had rod-like morphology in the other specimens. The rod-like crystalline phases' lengths were small, i.e. <20 μm, in the TiO_2 and BaO containing glass-ceramics, but small addition of ZrO_2 enhanced the length of crystalline phases to approximately 50 μm. MTT assay was used for cell proliferation assessment. The toxicity of glass-ceramic samples was assessed by seeding the osteosarcoma cells (MG63) on powder extracts for 7, 14 and 28 days. MTT results showed that glass-ceramic samples were almost equivalent concerning their in-vitro biological behavior.

INTRODUCTION

Fluorapatite $(Ca_5(PO4)_3F)$ containing glass-ceramics have attracted attention because of their compatibility with the natural apatite of the human bone and teeth[1,2]. Fluoroapatite is a bioactive compound with an apatite-like structure, in which OH groups have been substituted by fluorine ones. Furthermore, its crystalline structure is more stable than hydroxyapatite, which is an important issue where bioactivity receives attention based on crystalline stability[1-3]. Similarity between hardness of fluorapatite-based glass-ceramics and tooth enamel is another advantage that makes them new candidates for substitution of hydroxyapatite in restorative dentistry[1-3]. In most cases, bulk crystallization is the dominant mechanism of crystallization of needle- like fluorapatite in glass specimens. It is said that using P_2O_5 in the glass SiO_2-Al_2O_3-CaO-CaF_2 system leads to the crystallization of fluorapatite and mullite phases[4-9]. In these specimens fluorapatite is responsible to link to the tooth enamel and bone[3] and mullite induces adequate mechanical properties[10]. The latter phase plays a key role in biocompatible glass-ceramics since they are mostly planned to use in the stress bearing areas[2]. Despite of various studies that were performed on the chemical resistance, flexural strength, biocompatibility and phase separation of the apatite-mullite glass-ceramics[4-10], to our knowledge there are few documents about the performance of minor glass ingredients in points of crystallization behavior and biocompatibility properties. The primary purpose of the present study was to gain more insight to the effect of some components like TiO_2, ZrO_2, and BaO and also additional amount of silica in the mentioned regards.

MATERIALS and METHODS

The initial glass composition contained $4.5SiO_2$, $3Al_2O_3$, $1.5P_2O_5$, $3CaF_2$, and 2 CaO (mole ratio). Reagent grade chemicals Al $(OH)_3$, CaF_2, TiO_2, ZrO_2, $BaCO_3$, $CaCO_3$, phosphoric acid, and SiO_2 were used as the starting materials in this study. The chemical composition of the prepared glasses, which are coded as G, GS, GZ, GB and GT, are shown in Table 1.

Table 1. The chemical composition of different glasses

Glasses	G	GS	GZ	GB	GT
SiO_2	21.03	31.27	20.49	20.49	20.49
Al_2O_3	35.06	30.51	34.15	34.15	34.15
CaO	8.60	7.47	8.36	8.36	8.36
CaF_2	18.93	16.47	18.44	18.44	18.44
P_2O_5	16.38	14.26	15.96	15.96	15.96
TiO_2	----	----	----	----	2.56
BaO	----	----	----	2.56	----
ZrO_2	----	----	2.56	----	----

The thoroughly mixed batches were melted in an alumina crucible at 1550°C in an electric furnace for 2 h. Then the molten glasses were cast into a pre-heated steel mold. The resulting glass specimens were cooled naturally to room temperature. Glass transition temperatures and crystallization behavior were investigated using differential thermal analysis (Shimadzu DTG 60 AH). To study the crystallization behavior of the different glasses, the glasses were heat treated from 780°C to 1200°C at a heating rate of 10°C/min for 3 h and then furnace cooled to room temperature. The crystallinity of the specimens was identified by X-ray diffractometry (Jeol JDX-8030 and Siemens-D500). The microstructure of the samples was examined using a scanning electron microscope (Philips – XL 30 and Cambridge – S 360). The samples were etched by 10 wt. % HF solution prior to SEM analysis. MTT assay was used for cell proliferation assessment. The toxicity of glass-ceramic samples was assessed by seeding the osteosarcoma cells (MG63) on powder extracts for 7, 14 and 28 days. The proliferation and differentiation rates of the osteoblast-like cells were evaluated using extracted powder prepared according to ISO 1993–5 procedure. 0.1 g of powder samples with different compositions were incubated in 1 ml of culture medium. At the end of 7, 14 and 28 days, the mediums were collected for use in different cellular assays. Pure culture medium kept under similar conditions was used as a negative control sample. The proliferation rate of the osteoblast-like cells next to different powder extracts was determined by conducting the MTT (3-[4,5-dimethylthiazol-2-yl]-2,5-diphenyltetrazolium bromide) assay. This test is based on the fact that active cells convert the yellowish MTT to an insoluble purple formazan crystal.

RESULTS and DISCUSSION

Figure 1 depicts the DTA traces of various glasses. The crystallization peak temperatures (Tp_1 and Tp_2) of different glasses have been summarized in Table 2. Accordingly, there are two distinct crystallization peaks in every one. Besides, the glass transitions (T_g) and the crystallization peak temperatures have changed with the addition of the mentioned oxides. Based on these results, while SiO_2 and ZrO_2 increase the crystallization peak temperatures and their T_g, TiO_2 and BaO show contradictory trends. This behavior is attributed to rising of the glass viscosity with increasing of SiO_2 and ZrO_2, through reduction of non-bridging oxygen and /or high ionic field strength of these two oxides[11-15]. A contradictory explanation can be recommended for the effect of barium and titanium oxides, i.e. they increase the numbers of non-bridging oxygen in the glasses and hereby decrease the viscosity of the glasses[16-21]. Figure 2 shows the X-ray diffraction patterns of the glasses after heat treatment at different temperatures. Although it has not been presented here, obtained XRD results indicated that fluoroapatite appears after heat treatment at 780°C only in the glasses GT and GB, probably due to the lower viscosity of these two glasses at this temperature. This phase appears with lower intensities in G, GZ and GS at 870°C. Although the present authors were not able to detect a liquid-liquid phase separation, crystallization of the fluoroapatite as the first crystalline phase in the P^{+5} and Ca^{2+} rich areas of a liquid-liquid phase separated glass, has been reported previously for the same glass system[4-6].

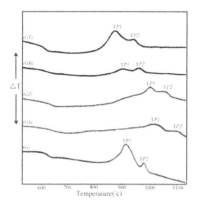

Figure. 1. DTA results of different glass

Table 2. Peak crystallization temperatures

Glass composition	TP1	TP2
GT	875 °C	950 °C
GB	900 °C	960 °C
GZ	1000 °C	≈ 1050 °C
GS	≈ 1015 °C	≈ 1105 °C
G	915 °C	975 °C

Mullite precipitates at 870 °C in GT and GB glasses and it is crystallized in GZ and G glasses at 970 °C. Based on the XRD results, apatite and mullite were precipitated in the glasses after the mentioned heat treatment and there was not any footprint of other crystalline compounds, such as TiO_2 and ZrO_2, in the XRD patterns, meaning that they did not act as a nucleation agent in the glasses. Furthermore, based on the DTA peak crystallization temperatures and the dilatometric softening point temperature of the glasses, it seems that GS has a higher viscosity than GZ. This makes the condition difficult kinetically for the precipitation of the crystalline phases in the GS glass composition and leads to reduction of their XRD peak intensities of the crystalline phases.

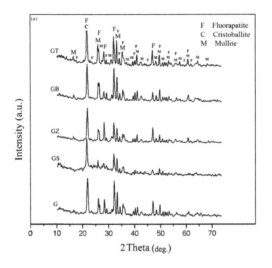

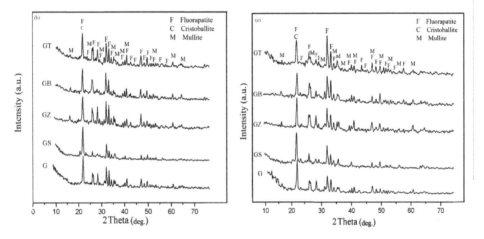

Figure. 2. XRD results of glass compositions heat treated at (a) 1100 °C and (b) 1200 °C for 3hr.

The microstructures of the glasses after heat treatment at 970 °C were investigated. Fig.

3a demonstrates the SEM micrograph of glass G after heat treatment at 970 °C for 3 h, respectively. This figure shows that heat treatment at 970 °C leads to precipitation of the rod-like crystals of apatite and mullite, with maximum length of about 5 μm. The morphology of crystals changes at 1100 °C. The crystalline particles dissolved in the glassy phase with the increasing of the heat treatment temperature to 1100 °C (pictures have been not presented here). The SEM micrograph of the heat-treated glass GS at 970 °C are shown in Fig. 3b, respectively. Accordingly, the spherical morphology of apatite and mullite crystals is independent of the heat treatment temperature. The microstructures of the glass-ceramics GZ and GB heat treated at 970 °C are shown in Figs. 3c and d, respectively. Based on these figures, the length of rod -like crystalline phases is between 5 -10 μm after heat treatment at 970°C. The crystalline particles kept their morphology even when the specimens were heated at 1100°C. Furthermore, the approximate lengths of crystalline phases increase to about 20 and 50 μm in GB and GZ glass-ceramics, respectively. Fig. 3e shows, the microstructure of glass-ceramic GT. Based on this figure, GT composition has also rod – like microstructure, with a maximum length of about 5 μm. The rod-like particles that according to EDX analysis could be attributed to both fluoroapatite and mullite crystals ultimately start to dissolve in residual glass phase at 1100 °C. The SEM micrographs of the samples support the XRD results. While apatite and mullite precipitate in the GS as spherical particles in the adopted heat treatment temperature interval, they precipitate in the other glasses as rod-like particles. This difference could be related to the high viscosity of the former glass, due to its higher SiO_2 content[21-23]. Spherical fluorapatite particles can also be seen in GT at lower temperatures. Based on the SEM figures, the size of rod shaped crystals in the glass-ceramics GB and GZ are significantly larger than that of G and GT and their aspect ratio has been increased with BaO and ZrO_2 addition. The mean lengths of crystals were approximately 20 and 50 μm in glass-ceramics GB and GZ, respectively.

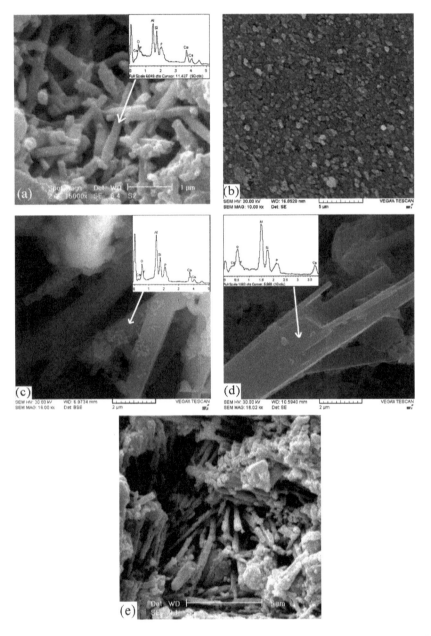

Figure. 3. SEM results of (a) G, (b) GS, (c) GZ, (d) GB and (e) GT compositions heat- treated at 970 °C for 3 h.

Figure 4 represents the results of the percentage cell viability for various powder extracts. MTT results (Fig.4) showed that glass-ceramic samples were almost equivalent concerning their in vitro biological behavior. The MTT results indicate that all of the glass-ceramic samples are biocompatible and the additive oxides and extra amount of SiO_2 don not led to the toxicity after 28 days cell proliferation.

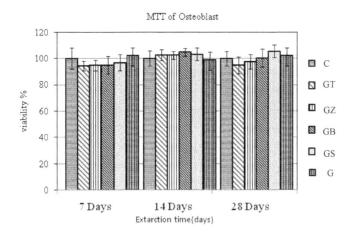

Figure.4. The osteoblast-like cell viability results for various extracted powder samples.

CONCLUSIONS

Addition of TiO_2, BaO, ZrO_2 and extra amounts of silica to the base glass led to some changes in the crystallization temperatures and morphology of the crystalline phases. While TiO_2 and BaO were effective in decreasing the crystallization temperature of the fluorapatite and mullite crystalline phases, ZrO_2 and the extra amounts of SiO_2 increased the crystallization temperature. Except for extra silica bearing specimen that showed spherical morphology for both apatite and mullite phases, apatite and mullite precipitated in the other prepared glass-ceramics as rod- like crystals. MTT results showed that glass-ceramic samples were biocompatible. Small addition of additives had a stimulating effect on the cell proliferation.

REFERENCES

[1] C. M. Gorman, R. G. Hill, Heat-pressed ionomer glass-ceramics. Part I: an investigation of flow and microstructure, *Dent. Maters* **19**, 320 –326 (2003).

[2] W. Holland, V. Rheinberger, S. Wegner, M. Frank, Needle – like apatite – leucite glass – ceramic as a base material for the veneering of metal restorations in dentistry, *J. Maters. Sci : Maters in Medicine* **11**, 11 – 17(2000).

[3] L. Montazeri, J. Javadpour, M. A. Shokrgozar, S. Bonakdar, S. Javadian, Hydrothermal synthesis and characterization of hydroxyapatite and fluorhydroxyapatite nano-size powders, *Biomedical Maters* **5**, 1– 8(2010).

[4] K. T. Stanton, R. G. Hill, Crystallisation in apatite-mullite glass–ceramics as a function of fluorine content, *J. Crystal Growth* **275**, e2061–e2068 (2005).

[5] K. Stanton, R. Hill, The role of fluorine in the devitrification of SiO_2–Al_2O_3–P_2O_5–CaO–CaF_2 glasses, *J. Maters. Sci* **35**, 1–6 (2000).

[6] M. D. O'Donnell, N. Karpukhina, A.I. Calver, R.V. Law, N. Bubb, A. Stamboulis, R.G. Hill, Real time neutron diffraction and solid state NMR of high strength apatite–mullite glass ceramic, *J. Non-Crys. Solids* **356**, 2693–2698 (2010).

[7] R. Hill, A. Calver, Real-Time nucleation and crystallization studies of a fluorapatite glass–ceramics using small-Angle neutron scattering and neutron diffraction, *J. Am. Ceram. Soc.* **90**, 763–768 (2007).

[8] K. T. Stanton, K. P. O'Flynn, S. Kiernan, J. Menuge, R. Hill, Spherulitic crystallization of apatite–mullite glass-ceramics: Mechanisms of formation and implications for fracture properties, *J. Non-Crys. Solids* **356**, 1802–1813(2010).

[9] A. Clifford, R. Hill, A. Rafferty, P. Mooney, D. Wood, B. Samuneva, S. Matsuya, The influence of calcium to phosphate ratio on the nucleation and crystallization of apatite glass-ceramics, *J. Maters. Sci. Maters in Medicine* **12**, 461– 469 (2001).

[10] C. M. Gormana, R. G. Hill, Heat-pressed ionomer glass–ceramics. Part II. Mechanical property evaluation, *Dent. Maters.* **20**, 252 – 261(2004).

[11] B. Eftekhari Yekta, P. Alizadeh, L. Rezazadeh, Synthesis of glass-ceramic glazes in the ZnO–Al_2O_3–SiO_2–ZrO_2 system, *J. Eur. Ceram. Soc.* **27**, 2311–2315(2007).

[12] M. Rezvani, The effect of complex nucleating agent on the physical and chemical properties of Li_2O-Al_2O_3-SiO_2 glass ceramic, *Iran. J. Mater. Sci. & Engin.* **7**, 8 – 16(2010).

[13] T. Wakasugi, R. Ota, Nucleation behavior of Na_2O-SiO_2 glasses with small amount of additive, *J. Non-Crys. Solids* **274**, 175–180 (2000).

[14] M. G. Garsia, J. L. Cuevas, C. A. Gutierrez, J. C. Angeles, J. F. Fuentes, Study of a mixed alkaline–earth effect on some properties of glasses of the CaO-MgO-Al$_2$O$_3$-SiO$_2$ System, *Boletin Sociedad Espanola de Ceramica* **46**, 153–162 (2007).

[15] Y. M. Sung, J. W Ahn, Sintering and crystallization of off-stoichiometric BaO-Al2O3-2SiO2 glasses, *J. Mater. Sci* **35**, 4913 – 4918 (2000).

[16] S. D. Matijasevic,V. D. Zivanovic, M. B. Tosic, S. R.Grujic, J. N. Stojanovic, J. D. Nikolic, S. V. Zdrale, Crystallization behaviour of Li$_2$O·Nb$_2$O$_5$·SiO$_2$ glass containing TiO2, Process. *Appl. Ceram.* **5**, 223–227(2011).

[17] E. S. Lim, B. S. Kim, J. H. Lee, J. J. Kim, Effect of BaO content on the sintering and physical properties of BaO–B$_2$O$_3$–SiO$_2$ glasses, *J. Non-Crys. Solids* **352**, 821–826 (2006).

[18] J. E. Shelby, Properties of alkali - alkaline earth metaphosphate glasses, *J. Non-Crys. solids* **263&264**, 271 – 276 (2000).

[19] S. Baghshahi, M.P. Brungs, C. C. Sorrell, H. S. Kim, Surface crystallization of rare-earth aluminosilicate glasses, *J. Non-Crys. Solids* **290**, 208 – 215(2001).

[20] P. F. Becher, M. J. Lance, M. K. Ferber, M. J. Hoffmann, R. L. Satet, The influence of Mg substitution for Al on the properties of SiMeRE oxynitride glasses, *J. Non-Crys. Solids* **333**, 124–128 (2004).

[21] Y. Zhang, J.D. Santos, Crystallization and microstructure analysis of calcium phosphate-based glass ceramics for biomedical applications, *J. Non-Crys. Solids* **272**, 14 – 21 (2000).

[22] Q. Xiang, Y. Liu, X. Sheng, X. Dan, Preparation of mica-based glass-ceramics with needle-like fluorapatite, *Dent. Maters.* **23**, 251–258 (2007).

[23] C. Moisescu, C. Jana, C. Rüssel, Crystallisation of rod-shaped fluoroapatite from glass melts in the system SiO$_2$–Al$_2$O$_3$–CaO–P$_2$O$_5$–Na$_2$O–K$_2$O–F⁻, *J. Non-Crys. Solids* **248**, 169–175 (1999).

Nanostructured Bioceramics and Ceramics for Biomedical Applications

SIZE CONTROL OF MAGNETITE NANOPARTICLES AND THEIR SURFACE MODIFICATON FOR HYPERTHERMIA APPLICATION

Eun-Hee Lee and Chang-Yeoul Kim

Nano-Convergence Intelligence Material Team, Korea Institute of Ceramic Eng. & Tech. 153-801 Seoul, Republic of Korea

ABSTRACT

Magnetite nanocrystals draw many attractions for their applications for hyperthermia therapy; cancer killing by heating. These days, hyperthermia together with chemotherapy shows improved cancer treatment effects. So, we tried to synthesize magnetite nanoparticles by thermal decomposition method with different sizes that are important magnetic properties. The size control of magnetite nanoparticles is controlled by changing a molar concentration of iron acetyl acetonate. We also focused on the surface modification of the magnetic nanoparticles with poly ethylene glycol (PEG) to give hydrophilicity and homogeneous dispersion within phosphoric acid-buffer solution (PBS) for the application of hyperthermia. We could present the possibilities of the dispersion of PEG-modified magnetite nanoparticles with different sizes and temperature increase under inductive magnetic field of 689Oe.

1. INTRODUCTION

Hyperthermia therapy is considered as one of promising cancer therapies with the well-known methods of surgery, chemotherapy and radiotherapy. There are two kinds of heating treatments: mild hyperthermia is performed between 41 and 46°C to stimulate the immune response for non-specific immunotherapy of cancers, and thermo-ablation at more than 46 up to 56 °C reported in the 1970s. Probably the oldest report was found in the Egyptian Edwin Smith surgical papers, dated around 3000BC. In the 19th and 20th centuries, fever therapy has been used as a method to increase temperature, while other investigators started to apply radiofrequency techniques [1]. In the past, external means of heat delivery were used such as ultrasonic or microwave treatments, but recently research has focused on the injection of magnetic fluids [2-4]. When a magnetic fluid is subjected to an alternating magnetic field, the particles become powerful heat sources, destroying tumor cell. The particles are preferably suspensions of superparamagnetic particles. The use of iron oxides in tumor heating was first proposed by Gilchrist et al. [5] and they investigated the hyperthermia effect of 20-100 nm-sized Fe_2O_3 particles under 1.2 MHz magnetic field. Since then, there have been numerous publications describing a variety of schemes using different types of magnetic materials, different field strengths and particles [6-13]. The principle of hyperthermia is that cancer cells are much more sensitive to and intolerant of the effects of excessive heat than normal cells. Also, tumours have an impaired ability to adapt their blood circulation to the effects of high temperatures and thus hyperthermia can cause a reduction of blood flow to a tumour. In addition, heat at this level decreases the viability and ability of spreading of cancer cells. The purpose of the study is to investigate the size control of magnetite nanoparticles and surface modification for the homogeneous dispersion in the human body fluid. We also study the relationship of the sizes and the magnetic properties and hyperthermia effects.

2. EXPERIMENT

2.1. Synthesis of magnetite NPs with iron(III) acetylacetonate concentration

1, 2, 3 and 4 mmol of iron(III) acetylacetonate (Fe(acac)$_3$, Aldrich Chemical) was dissolved into 20ml of benzyl ether, and 10 mmol 1,2 hexadecanediol, 6 mmol oleic acid and its equivalent mole of oleylamine were added. The key to successful synthesis of monodisperse magnetite NPs is preheating the solution at 200°C for 2h and then refluxing at 300°C for 1h under nitrogen gas flow (**Fig. 1**). Polyols like 1,2 hexadecanediol in this paper are more polar and form stronger associates with metal ions; chelating agents for metal ions to form metal complexes. Oleic acid is well-known to be capping ligand as a stabilizing agent. Iron chloride or nitrate reacts with oleic acid to form iron oleate. The thermal decomposition of the iron oleate is already researched to synthesize monodisperse magnetite NPs. In this study, we fixed 20ml benzyl ether as a solvent and 6mmol of oleic acid and equivalent molar concentration of oleyl amine were used as stabilizing agents and 1,2 hexadecanediol was used as a complexing agent. To separate magnetite NPs, we centrifugated the refluxed solution at 20,000rpm for 10 min and washed 3 times with ethanol and obtained oleic acid capped magnetite NPs. We changed molar concentrations of Fe(acac)$_3$ to investigate the relationships between the concentration and the particles sizes. (Table 1)

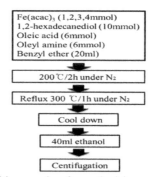

Fig.1. Schematic illustration of the procedure for synthesizing magnetite nanoparticles

Table 1. Chemical compositions and conditions for the synthesis of magnetite nanoparticles

Samples	Fe(acac)$_3$ (mmol)	Oleic acid (mmol)	Oleyl Amine (mmol)	1,2-Hexadec ane-diol (mmol)	Benzyl ether (ml)	Molar concentrati on (mM)
F1	1	6	6	10	20	40
F2	2	6	6	10	20	80
F3	3	6	6	10	20	120
F4	4	6	6	10	20	160

2.3. Surface modification of Magnetite NPs with molar concentration

The magnetic nanoparticles in biomedical application need to be hydrophilic for the homogeneous dispersion in human body and the introduction to tumor cells. Hydrophobic Fe$_3$O$_4$ colloid in hexane was transformed to hydrophilic Fe$_3$O$_4$ nanoparticles using polyethylene glycol (PEG). The solvent hexane was evaporated from the hexane dispersion of the particles under a flow of nitrogen gas, giving black solid residue of magnetite nanoparticles. The residue was dissolved in chloroform to form the chloroform dispersion at a concentration of 0.5 mg particles/mL solution. 1 mL of chloroform solution of polyethylene glycol (10 mg/mL) was added into 2 mL of the nanoparticle dispersion. The mixture was shaken at 37°C for 1 h and the chloroform solvent was evaporated under nitrogen gas in a vacuum oven for 1 d. The solid residue was dispersed in phosphate buffered saline (PBS) solution for further test.

2.4. Characterization of NPs and coating NPs

The crystal morphologies of magnetite nanoparticles were observed by transmission electron microscope (TEM; JEM 2000, 200 kV, point-to-point resolution 0.15 nm, JEOL Co., Ltd.). Fourier-transformed infrared (FT-IR, Prestige, Shimazu, Japan) spectroscopiess were also used to study the surface modification state of as-prepared magnetite nanoparticles and PEG-modified surfaces states. Magnetization (M)-magnetic field strength (H) relationships of magnetite NPs with different sizes were characterized by vibration sample magnetometer (VSM, VSM-5-10, TOEI IN. Co., Japan). Hyperthermia properties of magnetite NPs dispersed in water base was characterized by measuring temperature increase under the magnetic induction field generated by 2 turns copper coil where alternative output current of 80A at 266 kHz by magnetic induction heating equipment (PSTEK, Co. Korea). The samples were placed in the middle of coil and we measured the temperature changes with time by IR thermometer.

3. RESULTS AND DISCUSSION

3.1. Control of particle size with molar concentration

The crystal morphologies of magnetite NPs synthesized by the thermal decomposition were observed by TEM. Figure 2 shows TEM image of as-prepared magnetite NPs with molar concentration, which shows the crystal sizes increased with the molar concentration. As shown in Fig. 2, the crystal morphologies of samples of the F1(a) and F2 (b) was observed round shapes and uniform particles, but the crystal morphologies of samples of F3 (c) and F4 (d) had a little cubic shape. The more molar concentration of Fe(acac)₃, the more particles size of magnetite nanoparticles. It is considered that nucleation and growth of crystals are affected the number of ferrous ions, that is, the molar concentrations. Supposed that the monodisperse nuclei formed at nucleation state of 200°C, the growth will increase with the increase of molar concentration. The average values of F1, F2, F3 and F4 were 4, 7, 9.5 and 12.5nm. Secondary particle size was measured by laser particle size measurement using Mie scattering method. The results of the secondary particle size are given in Fig. 3. The average values of F1, F2, F3 and F4 were 520, 60, 35 and 25nm. It shows that the lesser molar concentration, the larger secondary particle size. It is assumed that the smaller primary particles tend to form the larger aggregation for the decrease of surface energies of the particles. Therefore, we guess that the appropriate size exists to form the homogeneous dispersion of magnetite nanoparticles. In our cases, we think that F4 samples are better to be homogeneously distributed in human body fluid due to the lesser aggregation. We think that less than 6 magnetite nanoparticles with about 10-13nm aggregates to form 25 nm-sized secondary particles.

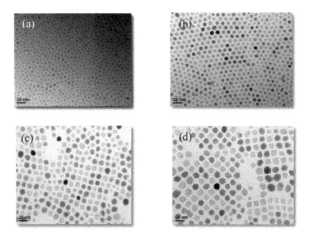

Fig. 2. TEM images of as-prepared magnetite NPs with different molar concentration, F1(a), F2(b), F3(c), and F4(d).

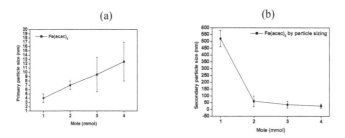

Fig. 3. The primary particle size variations (a) and the secondary particle size with different molar concentration (b).

3.2. Surface modification of magnetite nanoparticles

FT-IR spectra show the molecular vibration modes of as-prepared magnetite NPs with molar concentration and after washing (Fig. 4). As-prepared magnetite NPs showed peak of oleic acid at 2925 and 1735 cm^{-1} and peak of benzyl ether at 1500 cm^{-1}. It is known that oleic acid and solvent remained before washing. However, most of oleic acid and benzyl ether washed away from the surface of magnetite nanoparticles after washing. F4 samples showed stronger OH vibration band at around 3400cm^{-1}, which showed the more hydrophilicity. It indicates that washing away of oleic acid from the magnetite surface is more easily conducted from the larger particles. PEG surface modification of the washed magnetite nanoparticles confers the stronger hydrophilicity and the more stable dispersion in PBS solution. Our human body is comprised of water, so the dispersion of magnetite nanoparticles in PBS solution is very important. We could tell that the surface modification gives the compatibility of magnetite nanoparticles with viable cells.

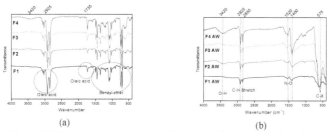

(a)

(b)

Fig. 4. FT-IR spectra of as-prepared (a) magnetite nanoparticles after washing (b).

3.3. Magnetic properties and hyperthermia effects

M-H characteristics of magnetite NPs with different molar concentration were measured by VSM. (Fig. 5) The response of the samples to external magnetic field strength, H, is typical of superparamagnetic characteristic behavior. It indicates magnetization (M) increases linearly at initial state and saturates gradually to some certain value. It did not show hysteresis loop due to single domain magnetite nanoparticles, because typical hysteresis appears due to multi-domains in a crystal grain. It is known that ferromagnetic to superparamagnetic transition occurs below 20 nm of particles, where they exist as single domains. The saturated magnetizations (Ms) of F1, F2, F3 and F4 were 40.7, 53, 58.7 and 60.7 emu/g, respectively. It indicates that magnetite nanoparticles with the larger sizes show the greater saturated magnetization values.

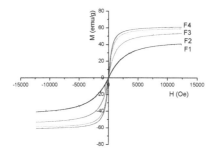

Fig. 5. M-H characteristics of magnetite NPs with molar concentration.

Hyperthermia properties of magnetite nanoparticles were analyzed indirectly by measuring the temperature changes of 10mg/ml concentrated magnetite solution in PBS in solenoid coils in two turns and 50 mm inner diameter under 80A at 266kHz. The AC magnetic field strength in the heating equipment was calculated by the magnetic field equation; $H = \frac{NI}{h}$, where H is the magnetic field magnitude in Oersteds, N is the number of coil turns, h is the heigth of the coil in meters and I is the current in amperes. The temperature change between initial and maximum temperature was 27 °C for F1, 30 °C for F2, 33°C for F and 34 °C for F4, as summarized in Table 2. It is thought that the lager magnetite nanoparticles generate heat more effectively. In temperature range of 42-45 °C, the function of many structural and enzymatic proteins within cells is modified, which in turn alters cell growth and differentiation and can induce apoptosis.

So, F4 samples increased the temperature to 50°C after 1h of AC induction heating, enough heat generation to kill cancer cells. The properties of magnetite NPs are summarized in table 2.

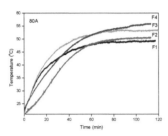

Fig. 6. Hyperthermia properties of magnetite nanoparticles encapped with PEG with different molar concentraton

Table 2. Properties of magnetite NPs with molar concentration

		F1	F2	F3	F4
Primary size (nm)		4	7	9	11
Secondary size (nm)		514.5	59.1	31.5	22.6
Yield (%)		53	77	86	91
Solid contents	as-prepared (%)	10	5	5	3
	After washing (%)	74.4	84	84.5	87
Ms (emu/g)		40.7	53	58.7	60.7
Hyperthermia	Max. temp. (°C)	49	51	54	56
	ΔT (°C)	27	30	33	34

4. CONCLUSIONS

We could control the sizes of magnetite nanoparticles from 4 to 13nm by varying the molar concentration. The aggregation of magnetite nanoparticles could be minimalized by the synthesis of magnetite nanoparticles with 13nm. The dispersion of magnetite nanoparticles could be conducted by washing oleic acid and surface modification with polyethylene glycol to confer hydrophilicity. The analyses of magnetic properties and the measurement of temperature changes under AC induction show that magnetite nanoparticles with larger size absorb the magnetic energy and generate heat more effectively to kill cancers upto 50°C for 1h of irradiation.

ACKNOWLEDGEMENTS

The authors are appreciated for the financial support of Korea Ministry of Trade, Industry and Energy as the core material research and development program (K0006028).

5. REFERENCES

[1] Seegenschmiedt MH, Vernon CC. A historical perspective on hyperthermia in oncology. In Seegenschmiedt MH, Fessenden P, Vernon CC (eds): Thermoradiotherapy and Thermochemotherapy Volume 1. Berlin: Springer Verlag 3-44 (1995)

[2] G.F. Baronzio, E.D. Hager, Hyperthermia in Cancer Treatment, Springer, New York, USA, (2006)

[3] S. Mornet, S. Vasseur, F. Grasset, E. Duguet, Magnetic nanoparticle design for medical diagnosis and therapy, J. Mater. Chem. 14, 2161-2175 (2004)

[4] M. Johannsen, B. Thiesen, A. Jordan, K. Taymoorian, U. Gneveckow, N. Waldofner, R. Scholz, M. Koch, M. Lein, K. Jung, S.A. Loening, Magnetic fluid hyperthermia (MFH) reduces prostate cancer growth in the orthotopic Dunning R3327 rat mode, The Prostate, 64, 283-292 (2005)

[5] R.K. Gilchrist, R. Medal, W.D. Shorey, R.C. Hanselman, J.C. Parrot, C.B. Taylor, Selective inductive heating of lymph nodes, Ann. Surg. 146, 596 (1957)

[6] M. Shinkai, M. Yanase, M. Suzuki, H. Honda, T. Wakabayashi, J. Yoshida, T. Kobayashi, Medical application of functionalized magnetic nanoparticles, J. Magn. Magn. Mater. 194, 176 (1999)

[7] A. Jordan, R. Scholz, P. Wust, H. Fahing, R. Feliz, Magnetic fluid hyperthermia (MFH): Cancer treatment with AC magnetic field induced excitation of biocompatible superparamagnetic nanoparticlesJ. Magn. Magn. Mater., 201, 413 (1999)

[8] T. Minamimura, H. Sato, S. Kasaoka, T. Saito, S. Ishizawa, S. Takemori, E. Tazawa, E. Tsukada, Tumor regression by inductive hyperthermia combined with hepatic embolization using dextran magnetite-incorporated microspheres in rats, Int. J. Oncol. 16, 1153 (2000)

[9] P. Moroz, S.K. Jones, J. Winter, A.N. Gray, Targeting liver tumors with hyperthermia: ferromagnetic embolization in a rabbit liver tumor model, J. Surg. Oncol. 78, 22 (2001)

[10] S.K. Jones, J.W. Winter, A.N. Gray, Treatment of experimental rabbit liver tumours by selectively targeted hyperthermia, Int. J. Hyperthermia 189, 117(2002)

[11] W.A. Kaiser, Efficacy of sequential use of superparamagnetic iron oxide and gadolinium in liver MR imaging, Aca. Radiol. 9, 198 (2002)

[12] Eun-Hee Lee, Chang-Yeoul Kim, Yong-Ho Choa Magnetite nanoparticles dispersed within nanoporous aerogels for hyperthermia application, Current Applied Physics, 12, 47-52 (2012)

[13] CY Kim, L Xu, EH Lee, YH Choa, Magnetic Silicone Composites with Uniform Nanoparticle Dispersion as a Biomedical Stent Coating for Hyperthermia, Advances in Polymer, 32, 723, (2012)

DESIGN, SYNTHESIS, AND EVALUATION OF POLYDOPAMINE-LACED GELATINOUS HYDROXYAPATITE NANOCOMPOSITES FOR ORTHOPEDIC APPLICATIONS.

Ching-Chang Ko*, DDS, MS, PhD
Department of Orthodontics, School of Dentistry, University of North Carolina, CB #7454, Chapel Hill, NC 27599, US
NC Oral Health Institute, University of North Carolina, Chapel Hill, NC 27599-7455, US

Zhengyan Wang, DDS, MS, PhD
Department of Biochemistry and Biophysics, University of North Carolina, Chapel Hill, NC

Henry Tseng, MD, PhD
Duke Eye Center and Duke University Medical Center, Durham, NC 27710

Dong Joon Lee, PhD
NC Oral Health Institute, University of North Carolina, Chapel Hill, NC 27599-7455, US

Camille Guez, DDS, MS
Department of Orthodontics, School of Dentistry, University of North Carolina, CB #7454, Chapel Hill, NC 27599, US

* Corresponding Authors: Ching-Chang Ko, Professor, Department of Orthodontics, University of North Carolina, Chapel Hill, NC 27599-7455. TEL: 919-537-3191; email: koc@dentistry.unc.edu

ABSTRACT

An ideal bone graft for restoring critical size defects (CSDs) should be sturdy enough to tolerate physiological loads, be malleable enough to fit into defects in the patient's bone, and stimulate bone regeneration. Here, we report a new nanocomposite material called Gemussel, made of gelatinous hydroxyapatite (HAP-GEL) with polydopamine crosslinking. We found that the mechanical performance of Gemussel is superior to calcium phosphate cement, can harden within aqueous environment, and is suitable for chair side applications. Gemussel is injectable and suitable for scaffolding. The combined physical properties of Gemussel were equivalent to that of cortical bone. Our result also showed that Gemussel had a compressive strength (120MPa), a biaxial flexure strength (110Mpa), and a compressive modulus (2.2GPa) that are approximating 80%, 50%, and 60% of cortical bone, respectively. We also found experimental evidence indicating that the polydopamine material in Gemussel may be bioactive and might play a role in stimulating bone formation. In support of this hypothesis, we found that osteoblasts express specific dopamine receptor subunits and adding polydopamine increased in vitro osteoblastic cell proliferation and differentiation. Mineralization and osteocalcin expression were increased by dopamine, suggesting an osteoconductive effect. To our knowledge, this is the first report which shows that osteoblasts express dopamine receptors and respond to dopamine. Taken together, our data demonstrate that polydopamine crosslinking significantly reduces brittleness of the previous HPA-GEL system and improves osteoconductivity possibly by stimulating bone formation through bioactive dopamine. Therefore, our new Gemussel bioceramics is an ideal material for restoring CSDs and bone regeneration.

INTRODUCTION

Critical size defects, defined as "the smallest size intraosseous wound in a particular bone that will not heal spontaneously during the patient's lifetime."[1,2], in bone can be difficult to manage and may require multiple-phase surgery with current graft materials to achieve adequate reparation and function.[3,4,5] An ideal bone graft should stimulate its replacement by the patient's own bone while being sturdy and formable to fit into defects. At present, the best graft material that can stimulate replacement by newly formed bone is the patient's own bone marrow or macerated bone, which has little or no structural strength or formability. Grafts of synthetic hydroxyl-apatite (HAP) are exceptionally resistant to loading but often are never replaced by the patient's own bone, and pose a long-term inflammatory risk.[6,7] Xenografts of tissues from other species can be strong initially but carry the risk of immune rejection and are replaced slowly.[8] Other current materials can be placed along a spectrum between the initial strength and replacability, compromising the ideal properties to provide balance between them. A graft material that provides high initial strength and good formability, but then resorbs and is coordinately replaced by the patient's own bone, is greatly needed. The purpose of this study is to report physical characteristics and osteogenic properties of the newly developed Gemussel nanocomposite.

Researchers have been always interested in natural products such as HAP, collagen (COL), and their mixtures for orthopedic applications because they mimic chemical elements of native bone. At present, none of the HAP-COL technologies can adequately address the three criteria (the initial strength, replacability, and formability). Built upon a HAP-gelatin (HAP-GEL) co-precipitation and sol-gel technology (an aminosilane cross-linker, enTMOS), we have recently reported an injectable bone composite graft, called Gemosil that can be used for direct scaffolding.[9] After it dries, Gemosil possesses a moderate compressive strength of 69 MPa. Nevertheless, the injectable mixture has high viscosity but low cohesive strength between HAP-GEL particles, and is prone to breakage and dissolution during gelation stage. The cohesive strength of the HAP-GEL particles relies upon two mechanisms: 1) interparticular electrostatic forces, van der Waals force, hydrogen bonds, and steric hindrance due to gelatinous polyampholytic and polyelectrolytic properties[10,11]; and 2) hydrogen bonds between siloxane and hydroxyapatite[12]. Although these bonds provide plasticity and processibility of Gemosil, weak bonding force only allows short distance sliding between the HAP-GEL particles. As a result, Gemosil is not suitable for indirect scaffolding such as salt leaching and computational topology design (CTD) because rehydration and any mechanical disturbance could disrupt the above mentioned bond formation. Applications of CTD-designed prostheses require an improved system that has fracture resistance (toughness) during molding and dissolution.

The present study introduces a tri-crosslinked Gemosil composite, called Gemussel, by employing two additional cross-linking reactions: 1) polydopamine via dopamine self-polymerization and 2) pozzolanic reaction that forms calcium silicate through the interaction between aminosilane and calcium hydroxide. Messersmith[13] discovered that dopamine, the adhesive proteins in mussels, can self-polymerize to form adherent polydopamine films for industrial use (e.g. coating[14], semiconductor, polymer[15]). We hypothesize that the co-existence of catechol and the amine group-mediated self-polymerization in dopamine can provide reactions similar to enTMOS for both chelation and gelation reactions in Gemosil. The pozzolanic reaction is the reaction between a pozzolan (a finely ground siliceous material) and $Ca(OH)_2$ in the presence of water to produce an insoluble calcium silicate hydrate (C-S-H).[16,17] In our application, enTMOS works as a pozzolan to react with $Ca(OH)_2$ to form a water-resistant moldable putty for scafflding. Both reactions significantly improved physical properties of Gemosil. The present report focuses on polydopamine rather than pozzolanic effect, which will be addressed in a companion publication. The compositional design, material processing, 3D scaffolding, and evaluation of Gemussel are presented herein.

Incorporating dopamine into the biomaterial offers further biological advantages. In cell cultures, Gemussel enhances osteoblast adhesion, osteoblast proliferation, and the rate of extracellular mineralization. Dopamine has been well-characterized as a neurotransmitter in many neurological processes and it is possible that osteoblast can respond to exogenous dopamine. Therefore, we hypothesize that Gemussel releases free dopamine which modulates osteoblast proliferation and differentiation. We found that osteoblasts express dopamine receptors 1, 3, and 5 (DrD1, DrD3, and DrD5), which supports our hypothesis and may provide a mechanism that explains the enhanced osteoblast response to this material. The role of dopamine in bone mineralization has not been explored. To our knowledge, this is the first report that osteoblasts express dopamine receptors and respond to dopamine. Our outcomes suggest a new era of osteo-neurobiology.

MATERIALS AND METHODS

Materials

The powder of hydroxapaptite-gelatin (HAP-GEL) was prepared by using the biomimetic process[18] followed by lyophilization process freeze drying[9]. Briefly, HAP-GEL slurry was synthesized by simultaneous titration of $Ca(OH)_2$ solution and an aqueous H_3PO_4 solution with a measured amount of gelatin (Type A : From porcine skin, Sigma-Aldrich, St. Louis, MO, USA) at 38°C and pH 8.0. The slurries were centrifuged at 5000 rpm under 4°C for 30 minutes to remove excess water. The condensed HAP-GEL was freeze-dried at -80°C overnight, lyophilized until dry, and ground into fine particles. $Ca(OH)_2$ was produced by [hydration of CaO at 285°C, where CaO was generated by the calcination of $CaCO_3$ up to 1200°C. Bis[3-(trimethoxysilyl)propyl]ethylenediamine (enTMOS) was purchased from Gelest, Inc (Morrisville, PA, USA). Phosphate buffered saline (pH = 7.4) was obtained from Sigma-Aldrich Co. Dopamine (3-hydroxytyramine hydrochloride, 99% beta-(3,4-dihydroxyphenyl)-ethylamin Hydrochloride 99%) was purchased from ACROS.

Sample Preparation and Conditions

Optimization: In this study, the HAp-Gel content and the amount of initiator, ammonium persulfate, were optimized. 50 mg, 100 mg, 150 mg, and 200 mg of HAP-GEL powder were compared to maximize compressive strength. The HAP-GEL powder was mixed with 200 mg $Ca(OH)_2$, which was determined in a companion publication, and ground into fine powders in a mortar. The enTMOS liquid (384 µL) and dopamine powder (40 mg) were blended to the above powder mixture on a cold (-20°C) glass slab. 40 µL of ammonium persulfate (10%) was added to trigger polydopamine formation; at this stage, the material became pasty and injectable. 40 µL of ammonium persulfate was chosen by testing various volumes of 40-, 60-, and 80-µL to optimize its reaction with 150 mg HAP-GEL. The paste was then removed from the cold slab and loaded into a syringe mold to form the solid samples at room temperature. The working time from loading the sample to the mold was approximately 3 minutes. After hardening, the samples were removed from the mold and air dried for 3 days prior to testing.

Fixed composition: The amount of HAP-GEL that had the highest compressive strength was 200mg, and this was used to fabricate biaxial bending samples (12.5 mm diameter by 2.6 mm thickness) following the same procedure described above. The dry samples were used to assess their applicability in the preformed scaffold for orthopedic applications. In contrast, samples without dopamine incorporation and samples without cold stage preparation were tested for comparison. Because of the injectability of Gemussel, we also measured both compression and biaxial bending to assess its mechanical strengths of samples that were immersed in the PBS for 24 hours after five minutes of hardening. For use in cell culture, coated 35 mm dishes were prepared by spin coating with 20 uL of the fixed composition in methanol solution. The coated dishes were allowed to dry for 3 days and before overnight UV sterilization.

Mechanical Testing

Mechanical testing consisted of compressive and biaxial bending tests. Cylindrical-shaped samples with a 1:2 ratio of diameter (3.5 mm) to height (7.0 mm) were prepared for compressive test and round disc samples for biaxial bending were fabricated. Both tests were performed on an Instron 4204 (Canton, MA, USA) with a cross-head speed of 0.5 mm/min. Biaxial bending test procedure for flexure strength was adopted from Ban and Anusavice[19]. The disk sample was supported on three stainless steel balls (3mm in diameter), which were equally spaced along a 6.5mm radius (r_s). A stainless steel piston (radius=r_p=1.5mm) pressed on the disc concentrically with the three balls (Fig. 1). The maximum force at failure (P) was determined. The flexure stress at failure (σ_t in MPa) was calculated using the following expressions: $\sigma_t = AP/t^2$ and $A = (3/4\pi)$ [2 (1+v) $\ln(r_s/r_o) + (1-v) (2 r_s^2 - r_o^2) / 2 (d/2)^2 + (1+v)$] where $r_o = (1.6 r_p^2 + t^2)^{1/2} - 0.675t$ and v is the Poisson's ratio of 0.3. Testworks 4 software (MTS. Inc) was used to analyze the data and the compressive strength was determined from the maximum strength value on the stress-strain curve. Toughness was calculated by integration of areas under the stress-strain curve of the compressive test. The results were analyzed by one-way ANOVA and followed by Tukey HDS comparison among the material groups. JMP Pro 9.0 (SAS Institution Inc., RTP, NC) was used for all statistical analysis.

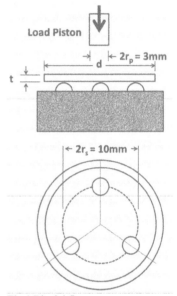

Fig.1 Biaxial flexure test apparatus

Transmission Electron Microscopy (TEM)

Transmission electron microscopy (TEM) was performed on a JEOL 2010F using an accelerated voltage of 200 kV. Samples were prepared by dispersing Gemussel powder in methanol and sonicating for 20 minutes. Then, formvar/carbon-coated TEM grids were immersed in the mixture solution followed by drying at room temperature.

Cell Culture Testing

Proliferation

35-mm dishes coated with Gemussel consisting of 200 mg HAp-Gel were used to study material's influence on preosteoblast proliferation and differentiation. Dishes coated with Gemussel without dopamine and uncoated dishes served as controls. Preosteoblasts, MC3T3-E1, were seeded at a density of 1×10^4 per milliliter using αMEM medium supplemented with 10% of FBS and 1% penicillin/streptomycin under 37 °C, 5% CO_2 atmosphere. Cellular proliferation activity was determined with MTS colorimetric assay at day 1, 4, and 7 according to the manufacturer's instructions (CellTiter96; Promega, Madison, WI, USA). Briefly, 40 µL of MTS reagent was added to the cells at the end of the designed cultivation period and incubated for 1 hour at 37 °C. One hundred µL solution were then taken out for measurement. Color formazan products were quantified by measuring absorbance at 490 nm (plate reader). The average background (medium plus MTS without cells) was subtracted from the absorbance values of the experimental and control groups for normalization purposes. The resultant optical densities were quantified to assess cell proliferation for each group. Three samples were tested with triplicate at each time point for each group. To assess cell growth morphology, we used coverslips that were coated with the Gemussle material, and seeded with MC3T3 cells in 35 mm dishes to examine cellular cytoskeleton and focal adhesion by phalloidin and vinculin immunofluorescence staining at day 3 after fixation and permeablization. Ccoverslips were then mounted on a slide by using antifade mounting solution. Fluorescence images were visualized with a fluorescence microscope (Nikon Eclipse Ti-U, Nikon Instruments, Melville, NY).

Differentiation
MC3T3-E1 was cultured on all three groups (Gemussel, Gemussel without dopamine, and the no-material control) in growth medium as described above for 3 day. Growth medium was then replaced with osteogenic medium (supplemented with 10 mM β-glycerophosphate and 0.2 mM ascorbic acid).. The osteogenic medium was changed on every third day. Assessment for cell differentiation was performed using Alizarin Red as a calcium mineralization stain. To understand whether the osteogenic effect was due to released substances of Gemussel, additional tests were conducted by adding the individual gelatin, $Ca(OH)_2$, or dopamine solution to the MC3T3-E1 culture and quantitative gene expression for type I collagen ($COL1\alpha2$), osteopontin (OPN), and osteocalcin (OCN) was evaluated on Day 1, 4 and 7. The cells were lysed for mRNA extraction using RNeasy Mini Kits (Qiagen). Reverse transcription for synthesis of cDNA was carried out using Omniscript Reverse Transcription Kit (Qiagen). Three samples for each group were tested and each sample was performed in triplicates. Real-time PCR reaction was accomplished by adding RT2 Qpcr Master Mix (RT2 SYBR Green/ROX qPCR Master Mix; SABiosciences) and primers to the first-strand cDNA synthesis reaction at room temperature, followed by RT-PCR in a 7500 Real-Time PCR System as described below. Three biological replicates were analyzed per sample condition. Expression level were quantitated relative to the control sample (which is untreated) and normalized to the mean expression of the housekeeping gene GAPDH.

Dopamine Receptor
We found that incorporation of dopamine into the substrate increased initial cell spreading, proliferation, and osteogenic gene expression. During culture, we also found that indeed, dopamine was released and accumulated in the culture media in Gemussel-coated culture dishes. Dopamine release was detected in the range 400nM (1hr) to 1µM (3days) by HPLC (in house unpublished data). To investigate whether dopamine receptors exist in osteoblasts, RT-PCR was used to screen all dopamine receptors and western blot was used to confirm the expression of the specific receptors of DrD1, DrD3, and DrD5. Immunostaining of dopamine receptor was also performed.

Reverse Transcription Polymerization Chain Reaction (RT PCR)
Briefly, total RNA was isolated from confluent differentiated MC3T3-E1 cells in each 35 mm dish by following instruction manual from Qiagen RNeasy Mini kits (Qiagen, Valencia, CA, USA), and then the RNA was reverse-transcribed into cDNA using an QuantiTect Reverse Transcription Kit (Qiagen, Valencia, CA, USA), allowing to compare the levels of gene

expression relative to the glyceraldehyde 3-phosphated dehydrogenase (GAPDH). Oligonucleotide primers for the PCR were designed for mouse dopamine receptors: Drd1a (F: 5'-ACCTACATGGCCCTTGGATGGC-3'; R: 5'-GGGAGCCAGCAGCACACGAA-3'), Drd2 (F: 5'-AGCCGCAGGAAGCTCTCCCA-3'; R: 5'-AGCTGCTGTGCAGGCAAGGG-3'), Drd3 (F: 5'-CCTGTCTGCGGCTGCATCCC-3'; R: 5'-TCTCCACTTGGCTCATCCC-3'), Drd4 (F: 5'-TCCTGCCGGTGGTAGTCGGG-3'; R: 5'-GCCAGCGCACTCTGCACACA-3'), Drd5 (F: 5'-TGGGAGGAGGGGCAGTCACC-3'; R: 5'-AGGTGGGCTCCTCCGTGAGC-3') and GAPDH (F: 5'-GCCACCCAGAAGACTGTGGAT-3'; R: 5'-TGGTCCAGGGTTTCTTACTCC-3') according to the reference.[1] The PCR conditions for the dopamine receptors and GAPDH were 29 cycles of denaturation (at 94°C for 40 s), annealing (at 55°C for 45 s), and extension (at 72°C for 40 s) followed by a final 5 min extension at 72°C. The PCR products were separated by electrophoresis through a 1% agarose gel containing GelRed Nucleic Acid Stain (Biotium, Inc., Hayward, CA USA) and image was capture by ImageQuant LAS 4000 (GE, Piscataway, NJ USA).

Western Blot Analysis

The differentiated MC3T3-E1 cells in 35 mm dish was washed with phosphate buffered saline (PBS), subsequently added 250 μL of lysis buffer (50 mM Tris-HCl, pH 8.0, 5 mM EDTA, 150 mM NaCl, 1% Triton X-100, 1 mM phenylmethylsulfonyl fluoride, and protease inhibitor cocktail) and stored at 4°C for 10 minutes. Cells were then scraped off and transferred to anEffendorf tube. The tube was kept on ice with frequent vortexing for 30 minutes to increase recovery of total proteins. The preparation was clarified by centrifugation, and the supernatant was saved as a whole-cell lysate. Total protein was measured by using Pierce BCA Protein Assay Kit (Thermo Fisher Scientific Inc., Rockford, IL USA), separated through 12% NuPAGE® SDS-PAGE Gel (Invitrogen, Carlsbad, CA USA) by different electrophoretic mobilities, and then electroblotted onto a nitrocellulose membrane (Millipore, Billerica, MA USA) by using Trans-Blot® SD Semi-Dry Transfer Cell (Bio-Rad, Hercules, CA USA). The membrane was then blocked with 5% nonfat dry milk in 25 mM Tris-HCl, 150 mM NaCl, and 0.2% Tween 20, and it was then incubated with the anti-dopamine receptor 1 and 3 antibodies (Millipore, Billerica, MA USA) and the anti-dopamine receptor 5 (Abcam®, Cambridge, MA USA) diluted 1:500 ratio using blocking solution for overnight. Each antibody was applied on different set of experiment. Subsequently, the membrane was washed and incubated for 1 hour with secondary antibodies conjugated to HRP (Milipore, Billerica, MA USA), rewashed, and developed using an enhanced chemiluminescence solution (Thermo Fisher Scientific Inc., Rockford, IL USA). Bands image was capture by ImageQuant LAS 4000 (GE, Piscataway, NJ USA).

Immunostaining of Dopamine Receptor 1a and 3

Localization of Dopamine receptor 1a and 3 on the MC3T3 preosteoblasts were detected by immunohistochemistry technique. MC3T3-E1 cells were cultured until approximately 50 to 60% confluent on the cover glass coated with type I collagen (5ug/ul). After washing three times with PBS, the cells were fixed with cold methanol for 30 minutes. Then, cells were exposed to blocking solution (10% FBS in PBS) for 30 min, and subsequently rinsed three times with PBS. Cells were incubated with the anti-dopamine receptor 1a and 3 antibodies (Millipore, Billerica, MA USA) diluted 1:250 ratio using 3% FBS solution at 4°C for 12 hours and then FITC conjugated secondary antibody (Millipore, Billerica, MA USA) diluted 1:500 ratio was applied at room temperature for 1 hour. The cells were rinsed three times with PBS for 30 minutes. Cell cultured cover glasses were mounted using a ProLong® Gold Antifade Reagent with DAPI kit (Molecular Probes/Invitrogen) on slide glasses. Images were acquired using a confocal microscope (Zeiss 510 Meta, Germany).

RESULT

Mechanical Characterization/Optimization
The amount of ammonium persulfate used significantly affected compressive strength of the Gemussel composites containing 150mg HAP-GEL (Fig.2A). The lowest amount of 40 µL was chosen for the remaining tests because it maximized compressive strength and minimized dehydration duration for sample fabrication. The amount of HAP-GEL significantly influenced compressive strength of the Gemussel (Fig.2B). Increasing HAP-GEL powders increased the compressive strength (P=0.02). Samples with HAP-GEL greater than 200mg resulted in a dry and powdery mixture, which hindered inject ability, and thus was not tested. The HAP-GEL not only improved compressive strength but also increased toughness due to its polydopamine contents. The plastic region of the stress-strain curve of the high HAP-GEL content was elongated far beyond the yield point (Fig.2C). The samples failed in several ways. With the higher strength samples it was likely to have failure with vertical cracks that sometimes shears a segment or segments completely from the sample. In general, the biaxial flexure strength of Gemussel reached 112.8±18.4 MPa, compared to the samples without dopamine incorporation that were extremely brittle and shattered into small pieces. The Gemussel samples that processed in room temperature without using the cold stage revealed short working time and one third of compressive strength (35MPa). The Gemussel could harden in an aqueous solution, unlike conventional calcium phosphate cement. To demonstrate this, the cylindrical samples mixed for two minutes and then immersed in PBS for 2 hours prior to testing. The compressive strength of the wet samples was around 12.2±1.2 MPa. TEM image shows that hydroxyapatite nanocrystal were bonded by polydopamine. The relatively small dopamine molecule (MW=153.18) is able to penetrate gelatin matrix of HAP-GEL particles and coat individual nanocrystalline HAP (Fig.3).

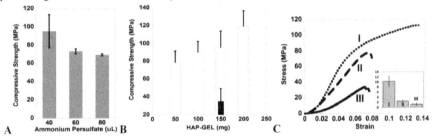

Fig. 2. A) Effect of ammonium persulfate on compressive strength of Gemussel. P=0.01; B) Effect of HAP-GEL contents on compressive strength of Gemussel with (white bar) and without (black bar) cold stage used. P=0.02; C) Typical stress-strain curves of compression tests and associated toughness for I: DA$^+$Gemussel, II: DA$^-$Gemussel, and III: Gemussel processed without using cold stage.

Biological Response

Cell Growth and differentiation
The MTS assessment showed that incorporation of dopamine in the substrate increased initial cell growth (Fig.4A). During the first 5 days, the cells were sporadic and did not spread on dopamine-negative Gemussel (DA$^-$Gemussel) while dopamine-positive Gemussel (DA$^+$Gemussel) appeared to resolve this problem. Fig.4B further confirmed this observation, showing relatively dense cell population for DA$^+$Gemussel, compared to DA$^-$Gemussel (Fig.4C)

at day 3. For differentiation assessment, alizarin red stain for calcium deposits increased when cells were cultured on both the DA⁻Gemussel and DA⁺Gemussel coated culture dishes, compared to the control (Fig.5A). The mineralization pattern appeared homogenous, fine crystals rather than conventional nodules. In SEM and TEM, the nano-apatite was closely associated with collagen that resembles in vivo bone ultrastructure[20], (Fig.5B). The analysis of how individual components leached from the Gemussel might affect cellular functions showed that both Ca(OH)₂ and dopamine increased cellular gene expression of OPN and OCN at days 4 and 7 (Fig.5C). The gene data indirectly support osteogenic property of Gemussel.

Fig. 3 TEM image shows HAP crystal (grey) bonded by polydopamine (white lines). Electric diffraction (upper right insert) confirms single crystalline HAP of the grey area.

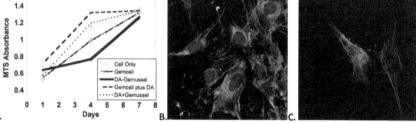

Fig.4. A: MTS assay showing cell growth curves were greater on dopamine (DA) doped materials. **B-C:** Confocal microscopy showing actin cytoskeleton and vinculin binding focal spots at Day 3. DA⁺Gemussel (B) shows high cell numbers with concentrated focal adhesions while DA⁻Gemussel (C) has low cell adhesion, confirming the absorbance data in (A).

Fig.5. (A) Alizarin red stain shows extracellular calcium deposition of MC3T3-E1 on DA⁻Gemussel (left 3 columns) and DA⁺Gemussel (the most right box) surfaces. Mineralization (darkness) increases over time. No-coating and no-cell (top row) do not show mineral stains before Day 7 of differentiation. (B) TEM image of the (A)* confirms apatite formation (black dots in collagen fibers) that blurs the 640Å band of type I collagen. The apatite is closely associated with collagen that resembles in vivo bone ultra-structures. SEM (top left insert) also confirms collagen fibers with mineral nodules. (C) Upregulation of osteogenic gene expression was further confirmed by treating cells with Ca(OH)₂ and Dopamine. The control is cell only without additional calcium, dopamine, or gelatin.

Dopamine Receptors

In PCR screening, three dopamine receptors (DrD1, DrD3, and DrD5) were detected but DrD2 and DrD4 were not (data not shown). Using western blot analysis, we further confirmed that osteoblasts (MC3T3-E1) expressed DrD1, DrD3, and DrD5 at protein level in differentiation medium at Day 7 (Fig.6A). This finding was also confirmed by immunofluorescence labeling (Fig.6B). To our knowledge, this is the first report that osteoblasts and rMSCs express dopamine receptors and respond to dopamine. These results suggest a possible mechanism of how Gemussel might promote bone mineralization through dopamine, which will be investigated in near future.

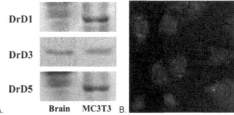

Fig.6 A: Western blot protein expression of DA receptors in differentiated MC3T3-E1 in Day 7. Mouse brain was used for the positive control. Dopamine receptors DrD1, DrD3, DrD5 are shown at 45-55kDa. **B.** Immunofluorescence stain of DrD1 (green) in MC3T3 (blue: nucleus).

DISCUSSION

Applying the self-assembly principle, we developed a biomimetic hydroxyapatite-gelatin nanocomposite (HAP-GEL) particles with preferred crystal orientation (along c-axis), which showed in vivo resorption potential by TRAP stain [21]. Each gelatin particle binds and immobilizes a cluster of nano-crystal HAPs that were precipitated in situ onto gelatin with - COO^-/Ca^{++} binding. One possible mechanism that results in resorbable property of HAP-GEL, unlike those seen in sintered HAP, may be attributed to cell binding of gelatin. In addition, the wetting ability of HAP-GEL particles significantly differs from conventional sintered HAP (Fig. 7). The HAP-GEL absorbs large amount of water while HAP does not, which gives HAP-GEL advantages for processing and results in greater mechanical strengths when it interacts with solvents (i.e., water and methanol).

Because particulate form of HAP-GEL is originally produced from the biomimetic aqueous reactor, efforts have been devoted to consolidate HAP-GEL particles into the solid composite and for scaffolding. Categorized by cross-linking agents used, the following three generations of formulation have been developed.

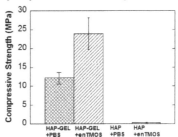

Fig.7. HAP-GEL with high wettability gives significantly greater compressive strength of the composite than that of sintered HAP when mixed with either PBS or enTMOS (siloxane linker). (n=5)

Generation One (G1)- Glutaraldehyde cross-linking: The first generation formulation used glutaraldehyde to cross-link HAP-GEL. The G1 exhibited compression strength ranging from 77 to 110 MPa, depending on glutaraldehyde content.[22] However, the yield of producing the testing samples was less than 10% due to its extremely brittle nature. It also took a long time (weeks) to dry. Glutaraldehyde binds to gelatin but not HAP, which resulted in a brittle nature of the final composite, cracks during dehydration, and breaks into pieces when immersed in water. No scaffolding could be produced from G1, which limits its clinical applicability. The G1, however, did show excellent in vitro and in vivo osteoconductivity[23] and resorption sign in eight weeks.[23]

Generation Two (G2) - Aminosilane cross-linking (Gemosil technology): To overcome the shortfalls of G1, aminosilane, N,N'-bis[(3-trimethoxysilyl)propyl]ethylenediamine (enTMOS) was selected as the cross-linking agent to create a new formulation (G2-Gemosil).[24,25] Gemosil, based on enTMOS sol-gel technology, consisted of HAP-GEL and siloxane. The Gemosil technology, using a powder form of freeze-dried HAP-GEL, controls the gel formation rate (≈working time) at room temperature through hydrolysis and polycondensation reactions to form a colloid (sol) that evolves then towards the formation of an inorganic network containing a liquid phase (gel).[26,27] The working time was 5-10 minutes, which allows shape molding[9]. FTIR showed that the hydrogen and Si-O-P bonding formed at the interface between the enTMOS matrices and HAP-GEL particles, which reduce the brittleness of the composite. The sample yielding rate is greater than 90% and the compressive strength of Gemosil was around 93 MPa. During the clay stage, the material is injectable to produce porous scaffolds. However, Gemosil cannot harden in water, which limits its clinical applicability.

Generation Three (G3) - Polydopamine cross-linking: Most recently, a third generation formulation (Gemussel) has been developed with adaptation of the mussel-inspired polydopamine cross-linker. Gemussel is built upon three cross-linking technologies: sol-gel process using enTMOS (forming siloxane), pozzolanic reaction[28] (Ca(OH)$_2$ reacts with enTMOS to form calcium silicate), and dopamine self-polymerization. Based on the present study, a specific proportion (Fig.8) of HAP-GEL, Ca(OH)$_2$, and dopamine/siloxane for Gemussel composites has been identified with a high compressive (120MPa) and flexure (110MPa) strengths. The optimal formula was composed of 39-43%wt HAP-GEL, 16-21%wt Ca(OH)$_2$, and 27-39%wt dopamine/siloxane(enTMOS) where the dopamine to enTMOS ratio was 1:15. Dopamine constituted only 5% weight of the total composite; however, the effect of polydopamine (black stain) was uniformly across the entire sample (Fig.9). Oxidation of dopamine and sequential polymerization was initiated by ammonium persulfate yielding black color change. In the present process, the polydopamine reaction occurred at -20°C, which allowed coating and kneading of HAP-GEL particles, while the sol-gel and pozzolanic reactions were halted. The kneaded paste then solidified (sol-gel and pozzolanic reactions) at room temperature (25°C). During the transition between -20°C and 25°C, the Gemussel paste could be molded to any desired shape. The mold ability can be used to fabricate a computer-aid scaffold as shown in Fig 9. Previous Gemosil is suitable for

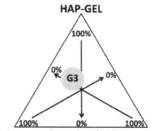

Fig.8. Area of triaxial composition shows optimal PDHG of 39-43%wt HAP-GEL, 16-21%wt Ca(OH)$_2$, and 27-39%wt dopamine/siloxane.

Fig. 9 Photograph of 3D Gemussel scaffold (700 μm pore) made of computer topology design indirect scaffolding.

direct scaffolding but not applicable for indirect computer-aid design. Gemussel has diverse applications as customized prostheses.

The mechanical performance of Gemussel approximated 80% and 50% of compressive and flexure strength of cortical bone, respectively. The HAP-GEL particles not only served as composite fillers to enhance compressive strength, but also increased toughness of the composite due to its gelatin content. The compressive modulus (linear slope of the stress-strain curve) was 2.2 GPa compared to 3.6 GPa of alveolar cortical bone of pig mandible. Adding polydopamine not only doubled flexure strength from 59MPa to 112MPa, but also increase 30% in compressive strength. The compressive stress-strain curve of Gemussel with dopamine depicts greater plasticity (or toughnes [29]) than that without polydopamine. Messersmith discovered that dopamine, the adhesive molecule in mussels, can self-polymerize to form adherent polydopamine films in water for industrial use (e.g. coating[30], semiconductor, polymer[31]). We hypothesized that the co-existence of catechol and the amine group-mediated self-polymerization in dopamine could substitute enTMOS for both chelation and gelation reactions in Gemosil. In fact, we found that the pasty mixture could harden in an aqueous environment with reduced strengths. In contrast, commercial calcium phosphate cement (CPC) cannot harden within aqueous environment. To assess the overall mechanical performance, we adapt a quality index (QI) previously proposed by William[32] to multiply compressive and flexure strengths and divide by compressive modulus. The modified QIs of CPC[33], HAP[34], and glass ionomer[35] are far inferior to natural bone [36, 37, 38] while Gemussel approximates natural bone (Table 1). The physical property of Gemussel approximates major criteria of ideal grafts being sturdy enough to accept physiological loading and formable enough to fit into bone defects.

Table 1. Physical properties (mean value) of Gemussel approximate natural bone. The quality index (QI= $\sigma_c \times \sigma_T / E^*$) modified from William's formula[32] includes σ_c, σ_T, and E^* for compressive strength, tensile strength (by biaxial flexure bending), and compressive modulus, respectively. GI= Glass Ionomer, and †= in house data. Gemussel has QI superior to commercial bioceramics and their hybrid materials.

Materials	σ_c MPa	σ_T MPa	E^* MPa	QI	Ref.
CPC	55	10	1000	0.55	33
HAP	509	80	20000	2.04	34
GI	120	38	900	5.07	35
Gemussel	120	112	2215	6.45	†
Cortical bone	150	175	3600	6.83	36-38

Incomplete autoxidation of polydopamine and noncovalent bonding interactions[39] could cause dopamine dissolution from a polydopamine matrix, making dopamine available to osteoblasts. Because dopamine is itself a neurotransmitter in the nervous system and another neurotransmitter called serotonin (5-HT) have been shown to regulate bone formation[40,41,42] and skeletal integrity[43], we hypothesized that dopamine may have osteogenic property through its interaction with dopamine receptors. In our *in vitro* experiments, it was clearly demonstrated that adding dopamine to the substrates increased osteoblast proliferation and profoundly increased mineralization stains. TEM further confirmed that the existence of fine apatite crystal associated with collagen fibers, which ruled out the concern that the red stain might be due to non-specific binding to free calcium. Another phenotypic effect was evidenced by increased gene expression in both osteocalcin (OCN) and osteopontin (OPN) due to dopamine treatment in differentiated osteoblasts. The present data provide a preliminary evidence that dopamine may directly affect both growth and differentiation of osteoblasts.

High dose (>10 μM) of dopamine was found to be detrimental to cell growth. In fact, we detected 400nM-1μM dopamine release in our culture dish (in house data), which was below the toxic dose. No adverse effect was observed in cell testing. Dopamine receptors that are supposed to respond to dopamine stimuli have been demonstrated in nervous, smooth muscle, and renal endothelial tissues, but not in bone. Which specific receptors are responsible for osteogenic activities is not known. Although the present study was directed to biomaterial design using polydopamine, we coincidentally discovered an osseous-neural connection. With positive outcomes in mechanical strength of Gemussel and evidence of dopamine receptors in osteoblasts, it necessitates further investigation in the era of Osteoneurobiology.

ACKNOWLEDGMENT
This study is supported, in part, by NIH/NIDCR K08DE018695, NC Biotech Grant#2008-MRG-1108, UNC Research Council, and American Association of Orthodontists Foundation.

REFERENCES

1. Schmitz J.P. and Hollinger J.O. (1986) The critical size defect as an experimental model for craniomandibulofacial nonunions. Clin Orthop Relat R 205: 299-308.
2. Hollinger J.O and Kleinschmidt (1990) The critical size defect as an experimental model to test bone repair methods. J Craniofac Surg 1: 60-68.
3. Gregory CF. (1972) The current status of bone and joint transplants. Clin Orthop Rel Res 87:165-166.
4. Losee JE and Kirschner RE. (2009) Comprehensive cleft care. McGraw-Hill Companies, Inc. pp. 837-988.
5. Sándor G.K.B., McGuire T.P., Ylikontiola L.P, Serlo W.S., Pirttiniemi P.M. (2007) Management of Facial Asymmetry; Oral Maxillofacial Surg Clin N Am 19:395-422.
6. Darimont GL., Cloots R, Heinen E, Seidel L, Legrand R. (2002) In vivo behaviour of hydroxyapatite coatings on titanium implants: a quantitative study in the rabbit. Biomaterials. 23(12):2569-2575.
7. Braye F, Irigaray JL, Jallot E, Oudadesse H, Weber G, Deschamps N, Deschamps C, Frayssinet P, Tourenne P, Tixier H, Terver S, Lefaivre J, Amirabaldi A. (1996) Resorption kinetics of osseous substitute: natural coral and synthetic hydroxyapatite. Biomaterials. 17(13): 1345-1350.
8. Alharkan A. (2007) Is Bio-Oss an osteoconductive material when used as an onlay bone substitute? An experimental study in the mandible of the rabbit. Faculty of Dentistry, M.Sc Thesis; McGill University.
9. Chiu C-K, Ferreira J, Luo T-J M, Geng H, Lin F-C, Ko CC. Direct Scaffolding of Biomimetic Hydroxyapatite-gelatin Nanocomposites using Aminosilane Cross-linker for Bone Regeneration. J Mater Sci: Mater Med. 2012; 23(9):2115-2126. PMCID:PMC3509178
10. Veis A. The macromolecular chemistry of gelatin. New York: Academic Press, 1964.
11. Darder M, Ruiz AI, Aranda P, Van Damme H, Ruiz-Hitzky E. Bio-nanohybrids based on layered inorganic solids: Gelatin nanocomposites. Curr. Nanosci. 2006;2:231.
12. Dupraz AMP, deWijn JR, vanderMeer SAT, deGroot K. Characterization of silane-treated hydroxyapatite powders for use as filler in biodegradable composites. J. Biomed. Mater. Res. 1996;30:231.
13. Lee H, Dellatore SM, Miller WM, Messersmith PB. (2007) Mussel-inspired surface chemistry for multifunctional coatings. Science. 318: 426-430.
14. Yang L, Phua SL, Teo JKH, Toh CL, Lau SK, Ma J, Lu X. (2011) A biomimetic approach to enhancing interfacial interactions: polydopamine-coated clay as reinforcement for epoxy resin. ACS Appl. Mater. Interfaces. 3 (8), pp 3026-3032.
15. Holten-Andersen N, Harrington MJ, Birkedal H, Lee BP, Messersmith PB, Lee KYC, Waite JH. (2011) pH-induced metal-ligand cross-links inspired by mussel yield self-healing polymer networks with near-covalent elastic moduli. *PNAS.* 108 (7): 2651-2655.

16. Jo B-W, Kim C-H, Tae G-H, Park J-B. (2007) Characteristics of cement mortar with nano-SiO2 particles. Construction and Building Materials. 21: 1351–1355.

17. Shi C and Day RL. (2000) Pozzolanic reaction in the presence of chemical activators. Part I. Reaction kinetics. Cement and Concrete Research 30:51-58.

18. Chang MC, Ko CC, Douglas WH. (2003) Preparation of hydroxyapatite-gelatin nanocomposite. Biomaterials. 24(17):2853-2862.

19. Ban S, Anusavice KJ. Influence of test method on failure stress of brittle dental materials. J Dent Res. 1990; 69(12):1791–1799. [PubMed: 2250083]

20. Olszta MJ, Cheng X, Jee SS, Kumar R, Kim Y-Y, Kaufman MJ, Douglas EP, Gower LB. (2007) Bone structure and formation: A new perspective. Materials Science and Engineering R 58:77-166.

21. Ko CC, Wu Y-L, Douglas WH, Narayanan R, Hu W-S. (2004) In Vitro And In Vivo Tests Of Hydroxyapatite-Gleatin Nanocomposites For Bone Regeneration: A Preliminary Report. In Biological and Bioinspired Materials and Devices. Edi. Aizenberg J, Landis WJ, Orme C and Wang R, Symposium Proceedings Material Research Society Spring Meeting. 823: 255-260.

22. Ko CC, Oyen M, Fallgatter AM, Hu W-S. (2006) Mechanical properties and cytocompatibility of biomimetic hydroxyapatite-gelatin nanocomposites. J. Material Research. 21(12):3090-3098.

23. Ko CC, Wu Y-L, Douglas WH, Narayanan R, Hu W-S. (2004) In Vitro And In Vivo Tests Of Hydroxyapatite-Gleatin Nanocomposites For Bone Regeneration: A Preliminary Report. In Biological and Bioinspired Materials and Devices. Edi. Aizenberg J, Landis WJ, Orme C and Wang R, Symposium Proceedings Material Research Society Spring Meeting. 823: 255-260.

24. Ko CC, Luo T-JM, Chi L, Ma A. (2008) Hydroxyapatite/gemosil nanocomposite. In Narayan R. and Colombo P. Eds. Advances in Bioceramics and Porous Ceramics: Ceramic Engineering and Science Proceedings. 29(7): 123-134.

25. Luo T-J M., Ko CC, Chiu C-K, Llyod J., Huh H. (2010) Aminosilane as an effective binder for hydroxyapatite-gelatin nanocomposites. J. Sol-Gel Sci & Tech. 53: 459–465.

26. Luo T-J M, Soong R, Lan E, Dunn B, Montemagno C. (2005) Photo-induced proton gradients and ATP biosynthesis produced by vesicles encapsulated in a silica matrix. Nature Materials. 4: 220-224.

27. Choi Y., Huh U. and Luo T. J. M. (2007) A Microfluidic Device for Opto-Electrochemical Sensing. NSTI-Nanotech, ISBN 1420061844, 3: 363-366.

28. Shi C and Day RL. (2000) Pozzolanic reaction in the presence of chemical activators. Part I. Reaction kinetics. Cement and Concrete Research 30:51-58.

29. Craig RG (1989) Restorative Dental Materials. 8th edi. Mosby, St. Louis. pp:85-86.

30. Yang L, Phua SL, Teo JKH, Toh CL, Lau SK, Ma J, Lu X. (2011) A biomimetic approach to enhancing interfacial interactions: polydopamine-coated clay as reinforcement for epoxy resin. ACS Appl. Mater. Interfaces. 3 (8), pp 3026–3032.

31. Holten-Andersen N, Harrington MJ, Birkedal H, Lee BP, Messersmith PB, Lee KYC, Waite JH. (2011) pH-induced metal-ligand cross-links inspired by mussel yield self-healing polymer networks with near-covalent elastic moduli. *PNAS.* 108 (7): 2651-2655.

32. William RM, Stephen EG, Gregory IB. (2001) Synthetic bone graft substitutes. ANZ J. Surg. 71:354–361.

33 Chow LC (2009) Next generation calcium phosphate-based biomaterials. Dental Materials Journal; 28(1):1-10.

34 Monmaturapoj N, Yatongchai C (2010) Effect of sintering on microstructure and properties of hydroxyapatite produced by different synthesizing methods. Journal of Metals, Materials and Minerals, 20(2):53-61.

35 Brantley WA and Eliades T. (2001) Orthodontic Materials- Scientific and Clinical Aspects. Thieme Stuttgart, New York. pp:229-236.

36 Yamada H (1970) Strength of biological materials. Williams & Wilkins, Baltimore.

37 Utz JC, Nelson S, O'Toole BJ, van Breukelen F. (2009) Bone strength is maintained after 8 months of inactivity in hibernating goldenmantled ground squirrels, Spermophilus lateralis. The Journal of Experimental Biology. 212:2746-2752.

38 Winwood K, Zioupos P, Currey JD, Cotton JR, Taylor M. (2006) Strain patterns during tensile, compressive, and shear fatigue of human cortical bone and implications for bone biomechanics. 79A(2):289–297.

39. Dreyer DR, Miller DJ, Freeman BD, Paul DR, Bielawski CW. (2012) Elucidating the structure of poly(dopamine). Langmuir. 28:6428-6435.

40. Westbroek I, van der Plas A, de Rooij KE, Klein-Nulend J, Nijweide PJ. (2001) Expression of Serotonin Receptors in Bone. J Biol Chem 276(31):28961–28968.

41. Bliziotes M, Gunness M, Eshleman A, Wiren K. (2002) The role of dopamine and serotonin in regulating bone mass and strength: Studies on dopamine and serotonin transporter null mice. J Musculoskel Neuron Interact. 2(3):291-295.

42. Rosen CJ. (2009) Serotonin Rising - The Bone, Brain, Bowel Connection. N Engl J Med 360(10):957-959.

43. Collet C, Schiltz C, Geoffroy V, Maroteaux L, Launay J-M, de Vernejoul M-C. (2008) The serotonin 5-HT2B receptor controls bone mass via osteoblast recruitment and proliferation. FASEB J. 22:418–427.

APPLICATION OF SCRATCH HARDNESS TESTS FOR EVALUATION OF PARTIALLY-SINTERED ZIRCONIA CAD/CAM BLOCKS FOR ALL-CERAMIC PROSTHESIS

Da-Jeong Lee[1], Seung-Won Seo[1], Hyung-Jun Yoon[1], Hye-Lee Kim[2], Jung-Suk Han[2], Dae-Joon Kim[1]

[1]Department of Advanced Materials Engineering, Sejong University, Seoul, Korea
[2]Department of Prosthodontics, School of Dentistry and Dental Research Institute, Seoul National University, Seoul, Korea

Correspondence to: D-J Kim; e-mail: djkim@sejong.ac.kr

ABSTRACT

Scratch hardness of partially sintered dental zirconia CAD/CAM blocks was determined using the load at which material removal by scratch occurs in a transient mode between the ductile and brittle manners. Scratch hardness increased at low indent loads and decreased at high indent loads. Specifically, the slope of the ascending linear relationship at low loads represents the extent of plastic deformation while the load at the beginning of the descending slope is related to crack opening by scratching, both of which govern machinability of the blocks. Here we demonstrate that the scratch hardness test reveals the machinability and shrinkage of CAD/CAM zirconia, two important pieces of information that may be utilized for fabrication of good dental prosthesis.

INTRODUCTION

One important facet in the recent upsurge of implementing ceramic technologies in the clinic has been the development of dental ceramics. Now a growing interest in dental aesthetics is taking on the form of tooth restoration, leading to a renewed demand for ceramics with high translucency and high strength.

Currently, three major types of ceramics are utilized for dental procedures: aluminosilicate glass, lithium disilicate-containing glass-ceramics, and polycrystalline zirconia.[1] Among the ceramics, 3 mol% yttria stabilized tetragonal zirconia, 3Y-TZP, is considered the most promising. It displays good biocompatibility, high fracture toughness and strength, and moderate aesthetics. However, a completely dense 3Y-TZP is not useful, as the hardness makes it difficult to machine into specific restorations such as crowns and bridges using conventional CAD/CAM systems deployed in most dental laboratories. As a solution to this problem, a partially sintered form of 3Y-TZP was recently adapted as a feeding block for dental CAD/CAM systems.

These partially sintered forms are initially machined 'over-sized', to compensate for the shrinkage of about 20% that occurs when subjected to a final sintering, in consideration of the accuracy of the final restoration fit. To compound the matter, recent evaluation of the as-sintered zirconia after the CAD/CAM machining of the partially sintered prostheses revealed a smear layer of flakes and wear debris coupled with extensive microcracking,[2] which can be detrimental to the quality of restorations. Poor restoration fit, coupled with irregularities in the surface caused by cutting paths[3] are major causes of poor marginal adaptation of restorations, all of which can increase risk of plaque retention leading to secondary caries, periodontal disease, and through microleakage, contribute to endodontic inflammation.[4] Thus, the improvement of the machinability and the precise control of shrinkage are important for the reliability of the restorations.

The machinability of the starting material has become an important consideration in selecting dental ceramics, since cutting and grinding are not only essential processes in dental laboratories but also the most commonly employed operations in rapidly advancing dental CAD/CAM systems. In this case, machinability can be defined as the relative ease or difficulty of removing material when transforming a raw material into a finished product which is further related to the ratio of fracture toughness to hardness.[5-7] Machinability of ceramics results from a chip-forming processs[8] so that the cut must be restricted to the vicinity of the cutting tool by a crack deflection mechanism for machinable dental ceramics. An easiness of machining may require a low hardness of ceramics and localization of cutting may be achieved by high fracture toughness. The machinability of ceramics has been evaluated by various test procedures, the scratch test being one of them.[9] Nevertheless, the evaluations have been restricted to fully sintered ceramics and no relevant studies have been reported for partially sintered CAD/CAM dental zirconia. In the present paper scratch hardness of commercially available dental zirconia blocks was evaluated in an effort to relate the hardness to the machinability and the shrinkage.

EXPERIMENTS

Five commercially available zirconia blocks for dental CAD/CAM systems was used in this investigation: Pearl NP (Acucera, Kyunggi-do, Korea), Pearl TR (Acucera, Kyunggi-do, Korea), Wieland (Wieland, Bamberg, England), Zirkonzahn (Zirkonzahn, Italy), and e.max CAD (ivoclar vivadent, Schaan, Liechtenstein). The partially sintered blocks were polished with 2000 and 4000-grits of SiC abrasives in sequence prior to scratch tests in accordance with ASTM G171-03. The scratch velocity was 15 mm/min and the loads were varied from 0.2 to 4.8 kg. After each scratch, the scratch tracks were observed using an optical microscope and field emission scanning electron microscope. The scratch hardness was calculated by

$$H_S = \frac{8P}{\pi W^2} \tag{1}$$

where H_S, P, $8/\pi$, and W are scratch hardness in GPa, normal force in Newton, geometric constant of diamond indenter, and scratch width in m. The widths at 9 different points randomly distributed along scratch distance of 10 mm were measured for each scratch, and the scratch hardness was determined from the average of 3 independent scratches for each specimen. The standard deviation of the measurements was less than 0.04 GPa. The tested blocks were sintered for 2 h at 1500 °C and the shrinkage was calculated by dimensional changes occurred during the sintering.

RESULTS AND DISCUSSION

A typical scratch hardness behavior is shown in Fig. 1. The hardness, determined using Eq.(1), increases with increasing loads (region I) and becomes constant (region II) and then drops with further increase in the loads (region III). At the bottom of the grooves in region I , material removal by the scratches was achieved by a plastic deformation in the ascending period. In Fig. 2 the scale-like bands were formed by squashing of the removed material through stick-slip events, which was also observed after scratching of sintered alumina,[10] as a result of continuous loading-unloading of the indenter. The bands, which represent cracks running perpendicular to the direction of the indenter, remained discrete. In region II, the cracks changed to being continuous and the distance between the bands became wider with increase in the loads. In addition to the change in shape, sporadic smearing of the bands and chippings were formed between the bands in this load range, suggesting that the plateau region represents a transition

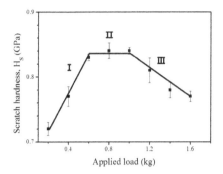

Figure 1. Applied loads dependence of scratch hardness of a partially sintered 3Y-TZP block, the Pearl NP.

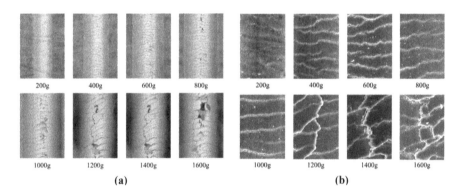

Figure 2. Morphology of scratch grooves of a partially sintered 3Y-TZP block, the Pearl NP, observed by (a) optical microscope and (b) SEM.

period in the material removal behaviors. In region III, a transverse mark, originating from the chipping, was observed across the bands. The transverse feature connected the bands to form a macroscopic defect running in the direction of the movement of the diamond indenter. A close examination revealed that the feature was formed by shearing of material in the direction parallel to the scratch and developed into cracks. Thus the load at the start of the descending hardness represents a critical point for crack opening. After observing that the hardness decreased under increasing loads, coinciding with the formation of the crack in region III, we determined that the descending trend of the hardness resulted from a brittle fracture of the partially sintered 3Y-TZP.

In the present study, the scratch hardness of the specimens was determined at the load entering region II because of termination of the indentation size effect, which is frequently observed in the Vickers hardness measurements of fully sintered ceramics.[11] The effect results

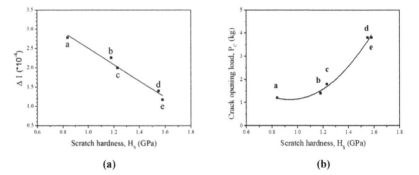

(a)   (b)

Figure 3. Scratch hardness dependence of scratch susceptibility (a) and crack resistance (b) of commercially available CAD/CAM zirconia blocks; (a) Pearl NP, (b) Pearl TR, (c) Wieland, (d) Zirkonzahn, and (e) e.max CAD.

from a recovery of plastically deformed material after unloading of the indenter. The hardness values of each specimen were related to their linear slopes, ΔI, in region I in Fig. 3(a) and crack opening loads, P_C, in Fig 3(b). The slope may represent the scratch susceptibility of the partially sintered CAD/CAM dental zirconia in such a way that the stiffer ΔI, the higher the resistance to scratching. For fully sintered silicon nitrides the susceptibility was determined by the scratch resistance measure that is proportional to the extent of volume removal and fracture strength and inversely related to the mean normal force, scratch length, and brittleness measure of the ceramics.[12] From the view of the size effect the slopes in Fig. 3(a) represent the recoverability of the deformed material and are plausibly related to the toughness that governs the machinability. On the other hand, the critical loads in Fig. 3(b) are associated with the chipping resistance as can be deduced from the scratch morphology shown in Fig. 2. Since a high ΔI is obtained from a block having low H_S and a high P_C is observed in a block with high H_S, compensation between ΔI and P_C is required for good machinability of the CAD/CAM zirconia blocks. Thus, a zirconia block having optimized values of ΔI and P_C is likely to be machined at relatively low load and the machined tracks are smooth because of less chance of the formation of macro-

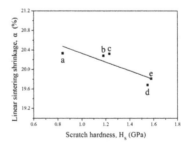

Figure 4. Correlation between shrinkage and scratch hardness of commercially available CAD/CAM zirconia blocks; (a) Pearl NP, (b) Pearl TR, (c) Wieland, (d) Zirkonzahn, and (e) e.max CAD.

cracks in the region III, accomplishing fabrication of excellent prosthesis.

Although the number of specimens and the extent of the shrinkage are limited, the scratch hardness is well correlated with the shrinkage of the specimens in Fig. 4. This is because material properties governing the hardness, such as packing density, pore size and distribution, and sintered density, also influence the shrinkage. The correlation may be valid only among the specimens sintered at either initial or intermediate stage of sintering because the material properties are completely different in the specimens sintered at each stage. Nevertheless, the present results demonstrate that the scratch hardness is a valuable property to predict the machinability and the shrinkage of the dental zirconia blocks. It is recommended that the scratch tests of the blocks, prepared at various sintering conditions, need to perform for the verification of the correlation.

ACKNOWLEDGEMENT

This work was supported by the National Research Foundation of Korea (NRF) grant funded by the Korea government (MSIP) (NRK 2010-0024260).

REFERENCES

[1]J.R. Kelly, Dental Ceramics: What Is This Stuff Anyway?, *J. Am. Dent. Assoc.*, **139**, 4S-7S (2008).

[2]J.W. Kim, N.S. Covel, P.C. Guess, E.D. Rekow, and Y. Zhang, Concerns of Hydrothermal Degradation in CAD/CAM Zirconia, *J. Dent. Res.*, **89**, 91-5 (2010).

[3]E.D. Rekow, N.R.F.A. Silva, P.G. Coelho, Y. Zhang, P. Guess, and V.P. Thompson, Performance of Dental Ceramics: Challenges for Improvements, *J. Dent. Res.*, **90**, 937-52 (2011).

[4]F. Beuer, H. Aggstaller, D. Edelhoff, W. Gernet, and J. Sorensen, Marginal and Internal Fits of Fixed Dental Prostheses Zirconia Retainers, *Dental Mater.*, **25**, 94-102 (2009).

[5]D.S. Baik, K.S. No, J.S. Chun, Y.J. Yoon, Mechanical Properties of Mica Glass-Ceramics, *J. Am. Ceram. Soc.*, **78**, 1217-22(1995).

[6]A.R. Boccaccini, Machinability and Brittleness of Glass-Ceramics, *J. Mater. Proc.Tech.*, **65**, 302-4(1997).

[7]X. Wang, G. Qiao, and Z. Jin, Fabrication of Machinable Silicon Carbide-Boron Nitride Ceramic Nanocomposites, *J. Am. Ceram. Soc.*, **87**, 565-70(2004).

[8]M. Taira, and M. Yamaki, Ranking Machinability of 9 Machinable Ceramics by Dental High-speed Cutting Tests, *J. Mater. Sci. Lett.*, **13**, 480-2 (1994).

[9]H.H.K. Xu, N.P. Padture, and S. Jahanmir, Effect of Microstructure on Material-Removal Mechanisms and Damage Tolerance in Abrasive Machining of Silicon Carbide, *J. Am. Ceram. Soc.*, **78**, 2443-8 (1995).

[10]Zhang Bi, H. Tokura, M. Yoshikawa, Study on Surface Cracking of Alumina Scratched by Single-Point Diamonds, *J. Mater. Sci.*, **23**, 3214-24(1988)

[11]J. Gong, J. Wu and Z. Guan, Examination of The Indentation Size Effect in Low-Load Vickers Hardness Testing of Ceramics, *J. Eur. Ceram. Soc.*, **19**, 2625-31(1999).

[12]G. Subhash, M. A. Marszalek, and S. Maiti, Sensitivity of Scratch Resistance to Grinding-Induced Damage Anisotropy in Silicon Nitride, *J. Am. Ceram. Soc.*, **89**, 2528-36(2006).

FUNCTIONALIZED ALKOXYSILANE MEDIATED SYNTHESIS OF NANO-MATERIALS AND THEIR APPLICATION

P. C. Pandey, Ashish K. Pandey
Department of Applied Chemistry
Indian Institute of Technology (Banaras Hindu University), Varanasi-221005

ABSTRACT

Organically functionalized alkoxysilane mediated synthesis of metal nanoparticles has been studied in detail. The experimental findings demonstrated the followings; (1) 3-Glycidoxypropyltrimethoxysilane (3-GPTMS) acts as efficient reducing agent and converts many metal salts like $AuCl_3$, $PdCl_2$, and $AgNO_3$ into their respective metal nanoparticles in the presence of 3-Aminopropyltrimethoxysilane (3-APTMS); (2) 3-APTMS acts as potent stabilizer, promotes the interaction of metal ions with 3-GPTMS, precisely controls the size of metal nanoparticles and also provides a suitable medium of nanoparticles suspension; (3) 3-APTMS significantly facilitates the stability of nanoparticles as compared to that made through conventional routes with common stabilizing agents. These nanoparticles illustrate functional activity for making nanocomposite and show size dependent enhancement in the electrocatalytic efficiency towards hydrogen peroxide and glutathione sensing. 3-APTMS also acts as an important reagent and precisely converts potassium hexacyanoferrate into Prussian blue nanoparticles in the presence of cyclohexanone.

I.INTRODUCTION

The synthesis of noble metals nanoparticles (NPs) is of immense interest due to their chemical, electrical and optica electro-catalytic properties.[1,2] Typical preparation of metal nanoparticles involves the reduction of a metal salt precursor solution in the presence of a reducing and stabilizing reagents and accordingly many reducing agents and stabilizers have been explored and reported in literature. However, many of them restrict their use in technological design especially in solid-state configuration due to aggregation phenomena. Accordingly, there still exits a need for the synthesis of noble metal nanoparticles while avoiding or controlling aggregation phenomena. A convenient way for introducing nanostructured domains in solid-state plateform has been very well studied in the synthesis and application of sol-gel materials derived from functionalized alkoxysilane precursors[4-6] where the relevant organic functionality allow to control the water wettable components during solid-state nanostructured network formation[7-11] In addition to that hydrophilic functionality like 3-APTMS has been used as stabilizers to complex gold ions and other metal ions in the sol form. whereas hydrophobic functionality, 3-GPTMS, allow the reduction of palladium chloride. Such findings directed our attention to investigate the role of organically functionalized moities especially 3-GPTMS and 3-APTMS from following anlges : (i) to understand the role of 3-GPTMS as reducing agent for metal salts, (ii) to understand the role of 3-APTMS during 3-GPTMS mediated reduction of metal salts, and (iii) to understand the advantages of functional ability of 3-APTMS in nanoparticle synthesis. Accordingly part of the present report demonstrate the contribution of 3-GPTMS and 3-APTMS as a reducing and stabilizing agents for the synthesis of noble metal nanoparticles of gold, silver and palladium which are amongst most required nanoparticles for practical applications and has been reported in this article. In addition to that, it is shown that the synthesized nanoparticles can be utilised for preparation of composite materials. Prussian Blue (PB) and nickel hexacyanoferrate (NiHCF) are chosen as a model system for composite preparation with above nanoparticles and a representative application of the latter is discussed here for the electrocatalytic determination of hydrogen peroxide and glutathione.

Further, the finding retaltes to reactivity of NH_2-group attached to alkoxysilane. Similar functional groups have been utilized for anchoring the targetted reactivity in specific molecules under ambient conditions.[12] Amongst many functionalities of organically functionalized silane the NH_2-functionality of 3-APTMS has been found as active centre especially for shiff-base formation in both solid-and liquid phase. In addition to that 3-APTMS also acts as potent stabilizer for nanoparticles. Such advantage of 3-APTMS has been examined in controlled synthesis of Prussian blue (PB) nanoparticles since, such synthesis has been challenging requirement due to lack of controlled nucleation of functional PB nanoparticles followed by simultaneous insertion of respective transition metal ion during nanoparticles nucleation and growth in three dimensional framework. We have observed novel finding on 3-APTMS and other organic moieties like cyclohexanone mediated controlled chemical synthesis of PB nanoaprticles. Thus, the finding on the role of 3-APTMS in stabilizing of as discribed nanoparticles and also to understand the functional reactivity of the same has also been reported in this investigation. Additionally a comparative study on the stabilization of gold nanparticles (AuNPs), made through conventional route of nanoparticle synthesis using common stabilizer and by 3-APTMS has also been observed and discussed in this article.

II.EXPERIMENTAL

Chemicals and Instrumentation:

The following chemicals were used: 3-aminopropyltrimethoxysilane (APTMS), 3-Glycidoxypropyltrimethoxysilane (GPTMS), were obtained from Aldrich Chem. Co. Tetrachloroauric acid, silver nitrate and palladium chloride were purchased from HiMedia; potassium ferricyanide, hydrogen peroxide, glutathione, and cyclohexanone were obtained from Merck, India. All other chemicals employed were of analytical grade. Aqueous solutions were prepared by using doubly distilled-deionized water (Alga water purification system).

The absorption spectra of samples were recorded in corresponding sol using a Hitachi U-2900 Spectrophotometer. Electrochemical experiments were performed on an Electrochemical Workstation Model CHI660B, CH Instruments Inc., TX, USA, in a three-electrode configuration with a working volume of 3 mL. An Ag/AgCl electrode (Orion, Beverly, MA, USA) and a platinum plate electrode served as reference and counter electrodes, respectively. The working electrode was a modified carbon paste electrode (CPE) having 2 mm diameter.

Preparation of silane stabilized AuNPs AgNPs and PdNPs:

Two sizes AuNPs were synthesized as follow: In a typical procedure,100 mL of sol was prepared having a molar ratio of 10 : 1 and 500:1 (APTMS : Au) respectively by mixing 8 mL solutions of 0.025 M of $HAuCl_4$ to 20 mL solutions of APTMS (0.05 and 0.5M respectively). Methanol was added as required to adjust the volume. The resulting yellow solution was stirred vigorously over a vertex cyclo mixer for 2 min. The solutions were then subjected to reduction by adding a 25 mL solution of 1 M GPTMS. The resulting mixture was stirred over a vertex cyclo mixer for 2 min. The mixture was left to stand in the dark for 12 h. After this, the colour of the sols turned to red and blue respectively indicating the formation of two different size of AuNPs. The method used to prepare silver nanoparticles was similar to that used for AuNPs. Typically, 20 mL sols of $AgNO_3$ were prepared containing APTMS: metal salt in the ratio of 5:1 and 20:1 in 99% methanol followed by addition of 5 mL solution of 1 M GPTMS. The suspension was homogenized and kept in dark for 12 h resulting into deep yellow colour sol of

AgNPs. Similar method has been used for the reduction of $PdCl_2$, the colour of sols turned to blackish in case of palladium nanoparticles (PdNPs).

Preparation of PB/AuNPs and NiHCF/AuNPs composite:

PB was synthesized as follows: 40 mL of 0.01 M aqueous solution of ferrous sulfate was added dropwise into a vigorously stirred solution of 40 mL aqueous solution of potassium ferricyanide (0.01 M).[13-15] The resulting deep blue solution was vigorously stirred for 5 min. 10 mL of the above solution was added to 10 mL of the above synthesized AuNPs solution (red and blue). The resulting mixture was then subjected to ultrasonication. To each of the above mixtures, 1 g of graphite powder was added to the above mixture followed by ultrasonication at 20 kHz for 20 min. The mixture was left to stand at 70^0C overnight for complete evaporation of solvent. The resulting solid residue (containing PB/AuNPs adsorbed on graphite powder) was collected and thereafter used for electrode preparation. Similarly, NiHCF was prepared following a single step protocol by adding 10 mL of an aqueous solution of nickle sulphate (0.01 M) to 10 mL of an aqueous solution of potassium ferricyanide (0.01 M). AuNPs of two different sizes was also allowed to interact and adsorbed on graphite powder similar as in case of PB.

Preparation of 3-APTMS mediated PBNPs:

The typical process of PBNPs synthesis involves the mixing of 50 mL aqueous solution of potassium ferricyanide (0.05 M) and 10 mL of 3-APTMS (0.5 M) under stirred conditions over a vertex cyclo mixer followed by the addition of 2 mL cyclohexanone (9.62 M). The mixture immediately turns into bright green color which subsequently converted to deep blue PBNPs sol after vigorous stirring. The resulting deep blue PBNPs sol was mixed with 5 mL HCl (6.5 M) under stirred condition.

Fabrication of modified carbon paste electrodes:

The as prepared PB/AuNPs or NiHCF/AuNPs adsorbed graphite powder was mixed in carbon paste electrodes (CPE) in the composition as PB/AuNPs or NiHCF/AuNPs = 2.5% (w/w), graphite powder = 67.5% (w/w), Nujol = 30% (w/w). The electrode body used for the construction of modified electrode was obtained from Bioanalytical Systems (West Lafayette, IN; (MF 2010)). The well was filled with an active paste of composition. The desired amounts of modifiers were thoroughly mixed with graphite powder in a blender followed by addition of Nujol oil. The paste surface was manually smoothened on a clean butter paper.

III.RESULTS AND DISCUSSION

Chemistry of 3-APTMS and metal salts interaction:

The interaction of 3-APTMS with differnt transition metal ions has been studied at first instant to understand the synthesis of nanoparticles mediated by 3-GPTMS and other similar agent. The visual photographs of 3-APTMS treated transition metal salts is shown in Fig.1A. The interaction of transition metal ions and 3-APTMS has been recorded by visual photographs in the presence (a) Mn(II), and (b) Cu(II) and absence (a') Mn(II) and (b') (CuII) of 3-APTMS (Fig.1A) that justify the capping of metal ions (lewis acid) with Amino-functionalities (lewis base) and also demonstrate that the affinity of such interaction increases with increase in charge to mass ratio of transtion metal ion based on similar finding recorded with orther trasition metal ions.

Another important finding of the role 3-APTMS is the reactivity of NH_2- functionality with [2-(3,4-Epoxycyclohexyl)ethyl]trimethoxysilane, allowing the control of hydrophobic and hydrophilic components, that leads into the formation of thin film of organically modified silicate.[5,12] These observations also predicts that 3-APTMS may facilitate the reducing ability of 3-APTMS-compatible reducing agent which is a key point during 3-GPTMS mediated synthesis of metal nanoparticles as the metal ions may opens the epoxide linkage of 3-GPTMS and the kinetics of such epoxide ring opening increases with increase in lewis acid character. The interaction of 3-APTMS with metal ions also justify reducing ability of 3-APTMS to metal ions under approprite reaction conditions having proton extraction sites during organic-inorganic hybrid interactions.

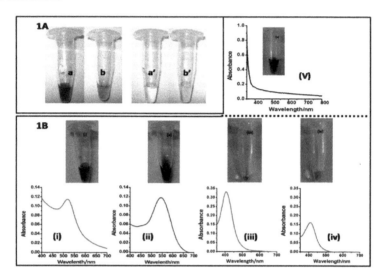

Fig.1 (A) The visual photographs of Manganese sulphate and Copper sulphate in 0.5M, 3-APTMS (a, b) and in water (a', b')respectively; (B) The visual photographs (upper portion) and corresponding UV-VIS spectra of AuNP$_{red}$ (i), AuNP$_{purple}$(ii), AgNPs of two size (iii-iv) and PdNPs (v) sol.

Chemistry of 3-GPTMS mediated reduction of metal salts in the presence of 3-APTMS:

In early 1990s, Schmidt and co-workers observed the interaction of glymo-residue of alkoxysilane with palladium chloride and concluded the reduction of palladium ion leading to the precipitation of Pd-glymo complex.[10] Later on, we demonstrated that GPTMS is highly sensitive to the presence of palladium chloride.[5,6] The epoxide ring of glymo-group is opened by palladium chloride followed by the reduction of palladium (II) into palladium. The reduced palladium is then coordinated within two moiety of glymo-residue.[5,6] Later on we further investigated the interaction of 3-GPTMS with several metal salt and recorded important finding on the reduction ability of the same. It was recorded that 3-APTMS play an important role during the reduction of metal ions. Fig.1B (i-v), show the photographs of gold nanoparticles (AuNPs) (i, ii); silver nanoparticles (AgNPs) (iii, iv) of two sizes; and ; Palladium nanoparticles (PdNPs) (v). The visual photgraphs are supported by UV-VIS spectra of the same justifying the synthesis of respective nanoparticles. The nanoparticles were further charcterized by Atomic force microscopy (Fig.2). AFM 2D and 3D images of AuNP$_s$ justify the average size of around 20 nm. The results also demonstrate (Fig.2A and B) that these nanoparticles are distributed within homogeneous silica matrix (the major part of images).

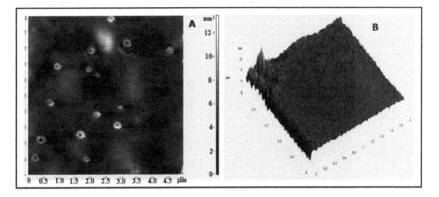

Fig. 2 AFM 2D (A) and 3D (B) image of AuNPs.

It is now important to understand the mechanistic role of 3-GPTMS mediated synthesis of nanoparticles in the presence of 3-APTMS. The epoxide linkage of 3-GPTMS may open in the presence of methanol and such epoxide ring opening leads to hydroxy and methoxy group functional 3-GPTMS. The interaction of 3-APTMS with metal ions justify reducing ability of 3-APTMS to metal ions under appropriate reaction conditions having proton extraction sites during organic-inorganic hybrid interactions.[12] Such finding reveals that 3-APTMS treated metal ions moves close to methoxy derivative of 3-GPTMS resulting into the formation of amine derivative through NH-group. The AuNPs formed thereafter are stabilized by 3-APTMS as well.

Applications of functional Nanoparticles made through 3-APTMS and 3-GPTMS mediated Synthesis:

The metal nanopartiocles synthesized mediated by 3-APTMS and 3-GPTMS are highly compatible and can be used to make nanocomposite, PB and NiHCF are chosen as a model system for composite preparation with above nanoparticles. A representative application of the same is discussed here for the size dependent electrocatalytic activity.

Size dependent electrocatalytic sensing of hydrogen peroxide:

First stage of investigation is to study the electrochemical behavior of PB/AuNP based modified electrode to understand the contribution of two different sizes of AuNPs on the redox electrochemistry of PB.[16, 17] The results based on cyclic voltammetry recorded for PB/AuNP$_{blue}$ and PB/AuNP$_{red}$ at different scan rate are shown in Fig. 3(A) and 3(B) respectively. The voltammograms clearly demonstrate gradual improvement in redox electrochemistry of modified electrode as a function of nanogeometry of AuNPs.

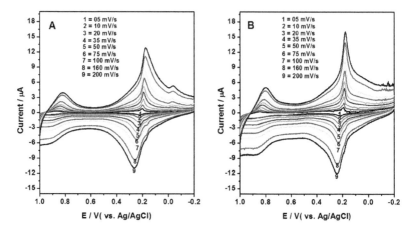

Fig. 3 Cyclic voltammograms of (A) PB/AuNP$_{blue}$ and (B) PB/AuNP$_{red}$ in 0.1 M phosphate buffer pH 7.0 containing 0.5 M KCl at various scan rates from 0.005 Vs^{-1} to 0.2 Vs^{-1}.

In order to understand the electrocatalytic behavior towards H$_2$O$_2$ sensing, the cyclic voltammogram of PB/AuNP$_{red}$ (Fig. 4a) and PB/AuNP$_{blue}$ (Fig. 4b) modified electrode with and without analyte have been recorded. The results showed that the cathodic current recorded on the addition of an identical concentration of H$_2$O$_2$ in the case of PB/AuNP$_{red}$ is relatively higher than that in the case of PB/AuNP$_{blue}$. For further support, amperometric measurement was conducted at 0.0 V and the results are presented in Fig. 4 C and D. It is clear from the results that in the case of PB/AuNP$_{red}$ modified electrodes higher current values are observed as compared to that of PB/AuNP$_{blue}$ modified electrode. The enhanced current value of the PB/AuNP$_{red}$ modified system indicates the role of AuNPs size in the composite material. The inset of Fig.4 C-D shows the calibration curve and the sensitivities towards H$_2$O$_2$ sensing was found to be 815 and 467 μAmM^{-1}cm^{-2} for PB/AuNP$_{red}$ and PB/AuNP$_{blue}$ modified electrode respectively.

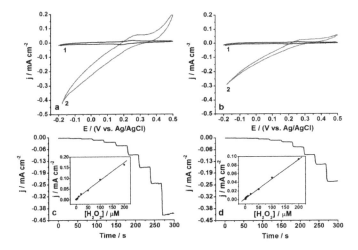

Fig. 4 (a) Cyclic voltammograms of PB/AuNP$_{Red}$ (a) and PB/AuNP$_{Blue}$ (b) modified carbon paste electrode in the absence (1) and the presence (2) of 5 mM H$_2$O$_2$ in 0.1 M phosphate buffer pH 7.0 containing 0.5 M KCl. Amperometric response of (c) PB/AuNP$_{Red}$ and (d) PB/AuNP$_{Blue}$ modified electrode on addition of varying concentrations of H$_2$O$_2$ between 0.01 mM to 5 mM at 0.0 V vs. Ag/AgCl. Insets show the corresponding calibration curves.

Size dependent electrocatalytic sensing of glutathione:

We further extend the investigation to evaluate the size dependent electrocatlytic ability of NiHCF nanocomposite with two different size AuNPs towards GSH sensing. The results based on cyclic voltammetry are shown in Fig.5 (A) and (B) indicate that the AuNP nanogeometry increases the catalytic efficiency of GSH analysis. These systems were used for quantitative determination of GSH based on linear sweep voltammetry (LSV). Fig. 5 C-D represents the LSVs of different concentrations of GSH at NiHCF/AuNP$_{blue}$ and NiHCF/AuNP$_{red}$ modified electrodes, respectively. The results justify that the presence of AuNPs fastens the reaction kinetics as a function of nanogeometry.

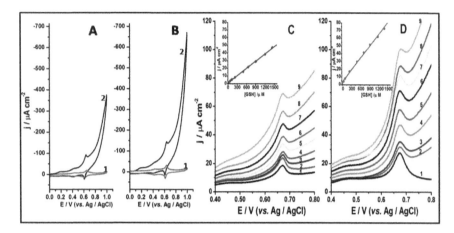

Fig.5 Cyclic voltammograms of NiHCF/AuNP$_{blue}$ (A) and NiHCF/AuNP$_{red}$ (B) modified electrode in the absence (1) and the presence (2) of 2 mM GSH in 0.1 M phthalate buffer pH 4.0. Linear sweep voltammetry of NiHCF/AuNP$_{blue}$ (C) and NiHCF/AuNP$_{red}$ (D) systems in 0.1 M phthalate buffer (pH 4) on the addition of 0.1, 10, 35, 50, 100, 250, 500, 700, 900, 1200, 1400 μM GSH (2-9). Insets show the corresponding calibration curves for GSH analysis.

We further examined the interaction of glutathione and AuNPs stabilized by different stabilizers i.e. polyvinylpyrrolidone (PVP) and 3-APTMS. Fig.6 shows the visual photographs and UV-VIS spectra of AuNPs stabilized by 3-APTMS (a, b) and by PVP (c) on the interaction of glutathione and found that 3-APTMS stabilized gold nanoparticles are relatively more stable (a, a' and b, b') as compared to that of PVP-stabilized AuNPs (c,c').Due to high affinity with metal nanoparticles and 3-APTMS, PVP-stabilized AuNPs are converted into blue color (Fig.6 c, c'). These finding clearly demonstrates the excellent capping ability of 3-APTMS with metal ions and their nanoparticles and direct potential advantage of the same in nanoparticles synthesis.

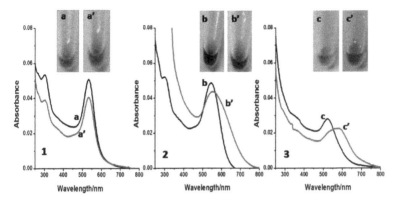

Fig.6 UV-VIS spectra and visual photographs of AuNPs in absence (a, b, c) and the presence of 0.1 mM GSH (a', b' c'); the recording shown as 1 and 2 (a, a'; b, b') represent the interaction of GSH and AuNPs made 0.05 M, (1) and 0.5 M, 3-APTMS (2) respectively whereas recording 3 (c, c') shows the interaction with PVP-stabilized AuNPs.

3-APTMS mediated synthesis of PBNPs

The functionalized alkoxysilane precursors are attracting from many angles. Out of their diverged utility, functionalized silanes have been used as stabilizers to complex metal ions in sol form.[7] These findings have directed the synthesis of of nanosized PB dispersions thereafter referred as PBNPs sol having variety of potential applications in the formation of functional nanocomposite sol, and mixed metal hexacyanoferrate nanoparticles sol. The role of 3-APTMS in controlling the size of nanoparticles (NPs) followed by acting as an stabilizer directed our attention to extract such ability for generating functional PBNPs sol. Indeed potassium hexacyanoferrate is found to display similar capping with 3-APTMS and provide ideal case of charge transfer complex formation in the presence of efficient electron donor and electron acceptor materials like 3-APTMS treated potassium hexacyanoferrate and cyclohexanone. The concentrations of 3-APTMS, potassium ferricyanide and cyclohexanone drastically affect the charge-transfer complexation leading to the conversion of PB nanoparticles. 3-APTMS further control the size of PBNPs followed by stabilization of same. The role of 3-APTMS is not only limited to such events and again provides a novel medium for the homogeneous dispersion of the PBNPs. An optimum concentration of each components i.e. potassium ferricyanide, 3-APTMS, and cyclohexanone, play central role for best PBNPs sol synthesis. Accordingly, the optimum concentrations of 3-APTMS are investigated as recorded in Fig.7A based on visual photographs and corresponding UV-VIS spectroscopy. The visual photographs and the respective absorptions spectra justify the results on the PBNPs sol conversion. Another important application is in the formation of nanocomposite sol with AuNPs. The AuNP synthesized via above described method is highly compatible with PBNPs sol and mixing of both sol resulted into the formation PB-AuNP nanocomposite under ambient conditions. The resulting PB-AuNPs nanocomposite sol justify as an excellent electrochromic and electrocatalytic functional nanomaterial for practical applications. The PB-AuNPs nanocomposite sol also justify its functional characteristics and enable the formation of excellent photochromic material with glutathione (GSH) Fig.7B shows the visual photographs and respective absorption spectra of PBNPs sol (1), PB-AuNPs nanocomposite sol (2) and reaction mixture of PB-AuNPs-GSH (3) and represent the multi functional behaviour of new material.

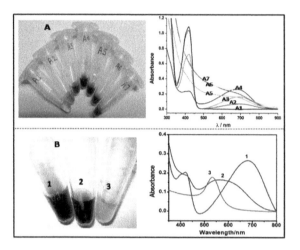

Fig.7 The photographs of: (A) systems containing constant concentrations of potassium ferricyanide and cyclohexanone followed with varying concentrations between 0.02 M to 2 M of 3-Aminopropyltimethoxysilane (A1 to A7) and the right portion shows the UV-VIS spectra of the corresponding sols. (B) The photographs of the PB (1), PB-AuNP(2),and PB-AuNP-GSH (3) sols, the right portion shows the UV-VIS spectra of the corresponding sols.

IV.CONCLUSION
The present work demonstrate the synthesis and applications of functional nanomaterials. The interaction of 3-APTMS with variety of transition metal ions is studied and the results justify the capping of metal ions (lewis acid) with Amino functionalities (lewis base) and also demonstrate that the affinity of such interaction increases with increase in charge to mass ratio of transition metal ion. It is found that 3-APTMS capped metal ions are converted into respective nanoparticles in the presence of 3-GPTMS. 3-APTMS facilitate the epoxide-group mediated reduction of metal ions followed by stabilization of the same and efficiently control the size of nanoparticles due to functional activity of the same during the reduction dynamics of the respective metal ions. A representative application of these nanoparticles reflecting size dependent electrocatalytic determination of H_2O_2 and glutathione is reported. Electrochemical sensing was greatly amplified as a function of AuNPs nanogeometry. The results based on interaction of glutathione and AuNPs stabilized by different stabilizers clearly demonstrates that 3-APTMS acts as powerful stabilizer as compared to that of others and represents excellent capping ability with nanoparticles. We also report the synthesis of PBNPs based on the interaction of active concentrations of 3-aminopropylalkoxysilane, cyclohexanone and potassium ferricyanide that allows the formation of functional nanocomposite with many noble metal materials, having excellent electrochromic and photochromic behaviors.

REFERENCES

1   K. Saha, S S. Agasti, C. Kim, X. Li, and V M. Rotello, Gold Nanoparticles in Chemical and Biological Sensing, *Chem. Rev.*, **112**, 2739-79 (2012).

2   A. Dass, R. Guo, J. B. Tracy, R. Balasubramanian, A. D. Douglas, and R.W. Murray, Gold Nanoparticles with Perfluorothiolate Ligands, *Langmuir*, **24**, 310-15 (2008).

3   P. A. Buining, B. M. Humbel, A. P. Philipse and A. J. Verkleij, Preparation of functional silane-stabilized gold colloids in the (sub)nanometer size range, *Langmuir,* **13**, 3921-26 (1997).

4   P. C. Pandey, and B. Singh, Library of electrocatalytic sites in nano-structured domains: Electrocatalysis of hydrogen peroxide, Biosens. Bioelectron., **24**, 842-48 (2008).

5   P. C. Pandey, S. Upadhyay and S. Sharma, Functionalized Ormosils-Based Biosensor Probing a Horseradish Peroxidase-Catalyzed Reaction, *J. Electrochem. Soc.*, **150**, H85-92 (2003).

6   P. C. Pandey, S. Upadhyay, I. Tiwari and S. Sharma, A Novel Ferrocene-Encapsulated Palladium-Linked Ormosil-Based Electrocatalytic Biosensor: The Role of the Reactive Functional Group, *Electroanalysis*, **13**, 1519-27 (2001).

7   L. M. Liz-Marzan, M. Giersig and P. Mulvaney, Synthesis of Nanosized Gold–Silica Core–Shell Particles, *Langmuir*, **12**, 4329-35 (1996).

8   B. Kutsch, O. Lyon, M. Schmitt, M. Mennig and H. Schmidt, Investigations of the electronic structure of nanoscaled gold-colloids in sol-gel-coatings, *J. Non-Cryst. Solids*, **217**, 143-54 (1997).

9   B. Kutsch, O. Lyon, M. Schmitt, M. Mennig and H. Schmidt, Small-Angle X-ray Scattering Experiments in Grazing Incidence on Sol-Gel Coatings Containing Nano-Scaled Gold Colloids: a New Technique for Investigating Thin Coatings and Films, *J. Appl.Crystallogr.*, **30**, 948-56 (1997).

10   T. Buckhart, M. Mennig, H. Schmidt and A. Licciulli, Nano Sized Pd Particles In A SiO$_2$ Matrix By Sol-Gel Processing, *MRS Proc.*, **346**, 779 (1994).

11   S. Bharathi, N. Fishelson and O. Lev, Direct Synthesis and Characterization of Gold and Other Noble Metal Nanodispersions in Sol–Gel-Derived Organically Modified Silicates, *Langmuir*, **15**, 1929-37 (1999).

12   A. P. Wight and M. E. Davis, Design and Preparation of Organic-Inorganic Hybrid Catalysts, *Chem. Rev.*, **102**, 3589-3614(2002)

13   P. C. Pandey, and A. K. Pandey, Electrochemical Behavior of Hydrogen Peroxide at Nanocomposite of Prussian Blue with Palladium of Variable Nanogeometery Modified Electrode, *J. Electrochem. Soc.*, **159**, G128-G136 (2012).

14   P. C. Pandey, and A. K. Pandey, Surface Modification Using Prussian Blue–Gold (I)–Palladium Nanocomposite: Towards Bioelectrocatalytic Probing of Hydrogen Peroxide, *BioNanoSci.*, **2** 127–134 (2012).

15   P. C. Pandey, A. K. Pandey and D.S. Chauhan, Nanocomposite of Prussian blue based sensor for l-cysteine: Synergetic effect of nanostructured gold and palladium on electrocatalysis, *Electrochimica acta*, **74**, 23-31(2012).

16   A. A Karyakin, E. E. Karyakina and L. Gorton, Prussian-Blue-based amperometric biosensors in flow-injection analysis, *Talanta,* **43**, 1597-1606 (1996).

17   A. A. Karyakin, O. V. Gitelmacher and E. E. Karyakina, Prussian Blue-Based First-Generation Biosensor. A Sensitive Amperometric Electrode for Glucose, *Anal. Chem.*, **67**, 2419-23 (1995).

DEVELOPMENT OF BIOACTIVE GLASS SCAFFOLDS FOR STRUCTURAL BONE REPAIR

Mohamed N. Rahaman[1,*], Xin Liu[1], and B. Sonny Bal[2]
[1]Missouri University of Science and Technology, Department of Materials Science and Engineering, Rolla, Missouri 65409, USA
[2]University of Missouri-Columbia, Department of Orthopaedic Surgery, Columbia, Missouri 65212, USA
*Corresponding author; e-mail: rahaman@mst.edu

ABSTRACT
The regeneration of large defects in load-bearing bones remains a clinical challenge. Current treatments such as bone autograft, allograft and porous metals have limitations. Bioactive glass is of interest in bone repair because it is bioactive, osteoconductive, converts to hydroxyapatite in vivo, and bonds strongly to hard and soft tissues. However, most previous studies indicated that bioactive glass scaffolds were sub-optimal for structural bone repair because of their low mechanical strength and concerns about their mechanical reliability in vivo. This article provides a review of our recent research in developing strong porous bioactive glass scaffolds to meet the need of repairing structural bone loss. Bioactive glass (13-93) scaffolds with a grid-like microstructure (porosity ~50%; pore width ~300 μm), created using a robotic deposition technique, have compressive strengths comparable to human cortical bone (100–150 MPa), good mechanical reliability in compression (Weibull modulus = 12), and excellent fatigue resistance (fatigue life >10^6 cycles) under compressive loads greater than normal physiological loads. Those strong porous scaffolds have shown the capacity to support new bone formation in osseous defects in a rat calvarial defect model. Eight weeks postimplantation, ~60% of the pore space in the scaffolds was infiltrated with new bone. Based on their mechanical properties and capacity to regenerate bone in osseous defects, these 13-93 bioactive glass scaffolds fabricated by robocasting are promising in structural bone repair.

INTRODUCTION
The repair of large defects in load-bearing bones, such as segmental defects in the long bones of the limbs, is a challenging clinical problem. A variety of commercially-available synthetic bone substitutes can be used to repair contained bone defects,[1,2] but none of them can replace structural bone loss. Traditionally, bone allograft, metal spacers, and even bone cement have been used to restore segmental bone defects in the limbs. These methods are limited by high costs, limited availability, unpredictable long-term durability, and uncertain healing to host bone. Millions of patients worldwide are affected each year by large bone loss resulting from civilian or war trauma, cancerous tumors, congenital diseases, and total joint surgery.[1] However, no product has successfully addressed the need for the replacement of structural bone loss in the limbs. Consequently, there is a need for new porous biocompatible implants that replicate the porosity, bioactivity, strength, and load-bearing ability of living bone.

In addition to being biocompatible, scaffolds intended for bone repair should be bioactive or bioresorbable. The scaffold should also have a porous microstructure suitable for supporting tissue ingrowth. An interconnected pore size (diameter or width of the openings between adjoining pores) of ~100 μm has been considered to be the minimum requirement for supporting tissue ingrowth,[3] but pores of size >300 μm may be required for enhanced bone ingrowth and formation of capillaries.[4] The scaffold should also have sufficient strength to withstand physiological loads. While there are no clear guidelines, it is generally thought that the as-fabricated scaffold should have mechanical properties at least comparable to the bone to be replaced. The strength of a bioactive or biodegradable scaffold intended for structural bone repair is sketched schematically as a function of implantation time in **Fig. 1**. As the scaffold degrades, the reduction in strength is offset by an increase in strength due to the ingrowth of new bone. The position of the crossover point is important. The strength should not be far below that of the bone to be replaced and the time should not be much longer than ~6 months for clinical applications. The design of scaffolds for skeletal reconstruction therefore involves an optimization of the mechanical properties of the scaffold and its capacity to support new bone growth.

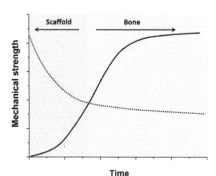

Fig. 1. Schematic diagram illustrating trends in the strength of bioactive glass scaffolds and new bone formation as a function of implantation time in vivo. The cross-over point between the two curves is arbitrary. In the first region (shorter implantation time), the scaffold provides the main load-bearing capability, while in the second region (longer implantation time), new bone formed in the scaffold should provide an increasing contribution to the implant strength.

Bioactive glasses, while brittle, have several attractive characteristics as a scaffold material for bone repair.[5-7] Bioactive glasses degrade chemically and convert to hydroxyapatite (HA), which bonds firmly to host bone. Calcium ions and soluble silicon released from silicate 45S5 bioactive glass promote osteogenesis and activate osteogenic gene expression. The compositional flexibility of glass can be used so that it is a source of many of the trace elements known to favor bone growth, such as boron, copper, and zinc; as the glass degrades in vivo these elements are released at a biologically acceptable rate.[8,9] Another advantage is the flexibility of preparing three-dimensional (3D) scaffolds with a wide range of architectures that could provide the requisite mechanical properties for load bearing, as well as an optimum physical and chemical environment for growing bone cells.

A variety of techniques have been used to fabricate bioactive glass scaffolds for bone repair, including thermal bonding of particles, spheres, or short fibers; consolidation of particles with a pore-producing fugitive phase; sol-gel methods; polymer foam replication; foaming of suspensions; freeze casting; and solid freeform fabrication.[6,10] The microstructure and the glass composition are the key factors that control the mechanical and biological performance of the bioactive glass scaffold. The microstructures produced by the methods described above cover a wide variety but, in general, they can be classified into three main groups (random; oriented; periodic) depending on the orientation of the pore (or solid) phase in the scaffold.

Examples of scaffolds in those three groups which were created from silicate 13-93 bioactive glass are shown in **Fig. 2**. Scaffolds created by thermally bonding a random packing of short glass fibers (100–200 μm in diameter × 3 mm in length) (**Fig. 2a**)[11] and by a polymer foam replication technique (**Fig. 2b**)[12,13] have a random arrangement of the pores. Unidirectional freezing of an aqueous or organic (camphene)-based suspension leads to the creation of scaffolds with a columnar microstructure of oriented pores (**Fig. 2c**).[14,15] Scaffolds with a periodic (grid-like) microstructure (**Fig. 2d**) were created using a robotic deposition or solid freeform fabrication techniques.[16-18] A key benefit of the robotic deposition techniques is the unprecedented control they can offer in creating pre-designed scaffold architectures.

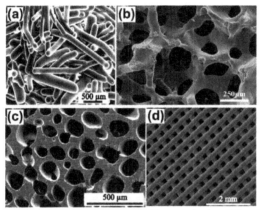

Fig. 2. Microstructures of bioactive glass (13-93) scaffolds created using different processing methods: (a) "fibrous" microstructure created by thermally bonding short fibers; (b) "trabecular" microstructure created by a polymer foam replication technique; (c) "oriented" microstructure formed by unidirectional freezing of suspensions (the pores are perpendicular to the planar section shown); (d) "grid-like" microstructure created by robotic deposition.

Several studies have evaluated the performance of bioactive glass scaffolds in vitro and in vivo, as reviewed elsewhere.[6,10] The main properties of interest in vitro are (1) the rate of conversion of the glass to HA, often used as a measure of the bioactivity (or bioactive potential) of the glass; (2) the response of the scaffolds to cells; and (3) the mechanical properties (strength; elastic modulus) of the scaffolds. The bioactivity and mechanical properties are commonly evaluated as a function of immersion time of the scaffold in an aqueous phosphate solution, such as simulated body fluid (SBF) that has a composition similar to that of human body fluid.[19]

Table I. Composition (in wt%) of silicate 13-93 glass used in this study, compared to the composition of silicate 45S5 glass and a borate glass composition designated 13-93B3, obtained by replacing the molar concentration of SiO_2 in 13-03 glass with B_2O_3.

Glass	SiO_2	B_2O_3	Na_2O	K_2O	MgO	CaO	P_2O_5
45S5	45.0		24.5			24.5	6.0
13-93	53.0		6.0	12.0	5.0	20.0	4.0
13-93B3		56.6	5.5	11.1	4.6	18.5	3.7

The bioactive glass composition has a strong effect on the bioactivity and the response of the scaffolds to cells. Silicate 13-93 glass (**Table I**) converts more slowly than silicate 45S5 glass while borate 13-93B3 glass converts more rapidly than 13-93 glass.[20,21] In vitro cell culture has shown no marked difference in the ability of 13-93 and 45S5 glass scaffolds to support the proliferation and function of osteoblastic cells.[22] In conventional (static) culture conditions, borate 13-93B3 glass was found to have a lower capacity to support the proliferation of osteogenic cells when compared to silicate 13-93 and 45S5 glass because of the high concentration of boron released from the glass into the medium.[23] However, under more dynamic conditions (e.g., periodic rocking of the culture system), the capacity of 13-93B3 glass to support cell proliferation was improved.

The mechanical properties of bioactive glass scaffolds are dependent primarily on the microstructure of the scaffolds, and secondarily on the glass composition. As fabricated, bioactive glasses show a "brittle" mechanical response in which the stress increases approximately linearly with deformation followed by catastrophic failure. The stress at failure is defined as the strength. **Figure 3** shows data compiled from the literature, in which the compressive strength of as-fabricated bioactive glass scaffolds are plotted as function of porosity of the scaffolds.[24] The data can be divided into three main regions that have a strong dependence on the microstructures described earlier. Scaffolds with a "random" microstructure typically have strengths in the range of trabecular bone (2–12 MPa), while scaffolds with an "oriented" microstructure show higher strength, with values between those for trabecular and cortical bone. In comparison, scaffolds with a "periodic" (grid-like) microstructure have shown the highest compressive strength, in the range of cortical bone (100–150 MPa).

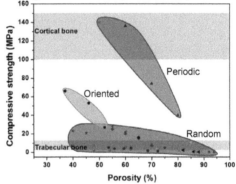

Fig. 3. Compressive strength of bioactive glass scaffolds compiled from data in the literature, and grouped based on their microstructure: random, oriented, and periodic. (From Ref. 24)

For the same microstructure, the compressive strength of as-fabricated borate 13-93B3 scaffolds was found to be lower than silicate 13-93 scaffolds, which was explained in terms of the lower strength of the B–O bond when compared to the Si–O bond.[18,25] Bioactive glass scaffolds, as described earlier, convert to HA when immersed in an aqueous phosphate solution, such SBF. Because the HA product is highly porous, the conversion process leads to a reduction in the strength of the scaffold. The rate at which the strength of the scaffold degrades is dependent on the glass composition. Borate glass (13-93B3) degrades faster than silicate (13-93) glass and, consequently, the strength of the 13-93B3 scaffold degrades more rapidly as a function of time in SBF. Because of their lower as-fabricated strength and rapid degradation rate, the use of bioactive borate glasses in structural bone repair might be challenging.

The composition and microstructure of bioactive glass scaffolds also have a strong effect on their ability to support bone ingrowth, mineralization, and angiogenesis and to heal osseous defects in vivo. After implantation for 12 weeks in rat calvarial defects, borate 13-93B3 scaffolds with a "fibrous" microstructure were found to have a better capacity to support bone regeneration than silicate 13-93 scaffolds with the same microstructure.[26] A greater amount of mineralization was associated with the 13-93B3 implants (new bone and conversion of the scaffold to HA). In another study, the effect of the scaffold microstructure on bone regeneration, mineralization, and angiogenesis was evaluated in a rat calvarial defect model.[27] Borate-based bioactive glass scaffolds with three different microstructures (fibrous; trabecular; oriented) were implanted for 12 weeks. Histomorphometric analysis showed that the borate 13-93B3 scaffolds with a trabecular microstructure had a greater capacity to support bone regeneration when compared to the oriented or fibrous scaffolds.

To summarize at this stage, previous studies using low-strength architectures showed that bioactive glass scaffolds were biocompatible in vivo, with the capacity to support bone infiltration. Recent studies showed that silicate bioactive glass scaffolds can be created with compressive strengths comparable to human cortical bone using robotic deposition techniques,[16–18] but characterization of those scaffolds was limited to their compressive strength and elastic modulus in vitro. In our recent work, we created strong porous scaffolds of silicate 13-93 bioactive glass by robocasting and comprehensively characterized their mechanical response in vitro and their capacity to regenerate bone in an osseous defect model in vivo. The results showed that those strong porous scaffolds 13-93 glass are promising in structural bone repair and they are briefly described in the following sections.

MECHANICAL PROPERTIES OF BIOACTIVE GLASS (13-93) SCAFFOLDS CREATED BY ROBOCASTING

Bioactive glass (13-93) scaffolds with a grid-like microstructure (**Fig. 4**) were created using a robocasting technique, as described elsewhere.[28] The scaffolds had a porosity of 47 $\pm$ 1%, a filament diameter of 330 $\pm$ 10 μm, and a pore width of 300 $\pm$ 10 μm in the plane of deposition (xy plane) and 150 $\pm$ 10 μm in the direction of deposition (z direction). The glass filaments were almost fully dense; the density as determined by the Archimedes method was 98 $\pm$ 1% of the density of the bulk glass (2.65 g/cm^3) formed by melting and casting. In addition, the glass filaments appeared to be well bonded to those in the adjacent plane.

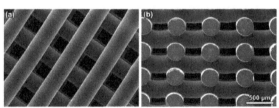

Fig. 4. SEM image of 13-93 bioactive glass scaffolds prepared by robocasting: (a) in the plane of deposition (xy plane); (b) in the direction of deposition (z direction).

The as-fabricated scaffolds were tested in compression and flexure using an Instron testing machine (Model 5881; Norwood, MA, USA). The compressive strength of scaffolds with a cubic shape (6 mm $\times$ 6 mm $\times$ 6 mm) was measured at a cross-head speed of 0.5 mm/min. The elastic modulus was determined from the linear region of the stress vs. strain response. The load was applied to the samples in the z direction of the as-formed scaffolds, perpendicular to the plane of deposition (xy plane) shown in **Fig. 4a**. This loading direction was used because it would match the compressive loading direction of the scaffolds in a segmental bone defect. Four-point flexural testing was performed on a fully articulated fixture (outer span = 20 mm; inner span = 10 mm) at a crosshead speed of 0.2 mm/min. The stress was applied in the z direction of the scaffolds (the same direction used for the compression tests). The as-fabricated scaffolds (3 mm $\times$ 5 mm $\times$ 25 mm) were tested according to the procedure described in ASTM C1674-11. Thirty samples each were tested in compression and in flexure, and the strength and modulus were expressed as a mean ± standard deviation (SD). The Weibull modulus in each loading mode was determined according to ASTM C1239-07.

The mechanical properties of the as-fabricated scaffolds in compression and in flexure are summarized in **Table II**. In compression, the scaffolds had a strength of 86 ± 9 MPa, a value that is close to the lower end of the values reported for human cortical bone. Modifications of this uniform grid-like microstructure, as described in our previous studies,[16,18] have resulted in compressive strengths similar to human cortical bone. The elastic modulus of the as-fabricated scaffolds was 13 ± 2 GPa, which is in the range of values reported for cortical bone. The flexural modulus of the scaffolds (13 ± 2 GPa) was in the range reported for cortical bone but the flexural strength (11 ± 3 MPa) was far lower than that of bone. However, the microstructure of the scaffolds has not been optimized yet. Current studies are aimed at re-designing the microstructure of the scaffolds to improve their flexural strength. Preliminary results indicate that scaffolds with a gradient microstructure designed to mimic bone, composed of an inner region with a microstructure shown in **Fig. 1a** (porosity ~50%) and an outer region with lower porosity (~20%), have flexural strengths that are twice the value given in **Table II**.

Table II. Mechanical properties of as-fabricated 13-93 bioactive glass scaffolds with a grid-like microstructure in compression and flexure (four-point bending). The load was applied in the z direction, perpendicular to the plane of deposition (xy plane) of the scaffolds.

Material	Compression			Flexure			Fracture toughness (MPa.m$^{1/2}$)
	Strength (MPa)	Elastic modulus (GPa)	Weibull modulus	Strength (MPa)	Flexural modulus (GPa)	Weibull modulus	
Scaffold	86 ± 9	13 ± 2	12	11 ± 3	13 ± 2	6	0.48 ± 0.04
Cortical bone[24,29]	100–150	10–20		135–193	9–16		2–12
Trabecular bone[24]	2–12	0.1–5		10–20			0.1–0.8

Commonly, the strength and Weibull modulus are used to evaluate the probability of failure of brittle materials under a given stress. The Weibull modulus of the scaffolds created in the present study was 12 and 6, respectively, in compression and in flexure. Data for the Weibull modulus of bioactive glasses are limited. The Weibull modulus of porous bioceramic scaffolds, such as HA, beta-tricalcium phosphate (β-TCP), and calcium polyphosphate (CPP), has been reported in the range 3–10 for testing in compression.[24,30-32] **Figure 5** shows a comparison of the Weibull plots for the bioactive glass scaffolds in this study with plots taken from the literature for calcium phosphate scaffolds tested in compression.[30,31] Under the same allowable failure probabilities, the bioactive glass scaffolds showed a compressive failure strength that is higher than that for the HA scaffolds, and far higher than that for the β-TCP scaffolds. Based on the strength and the Weibull modulus data, when subjected to a compressive stress of 50 MPa, the failure probability of the bioactive glass scaffolds, P_f, is equal to 10^{-3} (1 in 1000 scaffolds is predicted to fail). The average stress on a hip stem is reported as 3–11 MPa,[33,34] well below the stress (50 MPa) for a failure probability of 10^{-3}.

The fracture toughness of the as-fabricated scaffolds was measured by the single-edge notched beam (SENB) technique using samples of size 3 mm × 5 mm × 25 mm according to the procedure described in ASTM C1421-10. A notch of width <100 μm and depth of ~1.5 mm was machined at the midpoint of the 3 millimeter-wide plane. The notch depth (a few times the cell size of the scaffold) was chosen to satisfy the conditions for applicability of linear elastic fracture mechanics (K-dominance at the crack tip).[35] The fracture toughness K_C determined from testing five samples was $(0.48 \pm 0.04 \text{ MPa·m}^{1/2})$, which is much lower than the value for human cortical bone $(2-12 \text{ MPa·m}^{1/2})$,[36] but in the range of values for dense glass $(0.5-1 \text{ MPa·m}^{1/2})$, porous HA $(0.3 \text{ MPa·m}^{1/2})$,[37] and porous phosphate glass–ceramics $(0.2-0.6 \text{ MPa·m}^{1/2})$.[38]

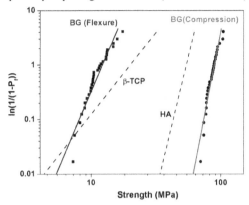

Fig. 5. Weibull plots of the compressive and flexural strength data from the present study for 13-93 bioactive glass scaffolds created by robocasting. For comparison, Weibull plots of the compressive strength data from the literature[30,31] for beta-tricalcium phosphate (β -TCP) and hydroxyapatite (HA) scaffolds fabricated by robocasting are also shown for comparison (dashed lines).

Fatigue testing of the as-fabricated scaffolds was performed in cyclic compression using an ElectroForce 3330 testing system (Bose Corp., Eden Prairie, MN, USA). Samples (6 mm × 6 mm × 6 mm) were tested in air at room temperature and in phosphate-buffered saline (PBS) at 37 °C using load-control actuation, at a frequency of 5 Hz. Three cyclic compressive stresses of 1–10 MPa, 2–20 MPa and 3–30 MPa were used, with the minimum to maximum stress ratio kept constant at 0.1. The tests were conducted until failure or until 10^6 cycles were reached. Six samples were tested at each cyclic stress. The fatigue life (mean ± SD) was determined using a logarithmic transformation of the cycles to failure as recommended in ASTM F2118-10.

The fatigue life of the as-fabricated scaffolds when tested in compression in air and in PBS is shown in **Fig. 6**. Of the 6 samples tested under each condition, the number that survived the 10^6 cycles of testing is also indicated. In air, all 6 samples tested under a cyclic stress of 1–10 MPa survived, showing that the samples had a fatigue life greater than 10^6 cycles. As the cyclic stress amplitude was increased to 2–20 MPa and 3–30 MPa, 4 and 3 samples, respectively, survived the 10^6 cycles of testing. Although the mean fatigue life decreased with the increase in the stress amplitude, the difference was not significant. Under a cyclic stress of 1–10 MPa, testing in PBS did not have a significant effect on the mean fatigue life when compared to the samples tested in air; 5 samples survived the 10^6 cycles of testing. However, testing under cyclic stresses of 2–20 MPa and 3–30 MPa resulted in a significant decrease in the mean fatigue life.

Because long limb bones undergo cyclic loading, the fatigue resistance of scaffolds designed for bone substitution is relevant. The data indicated that for the cyclic stresses used, the mean fatigue life was independent of the stress when the samples were tested in air, whereas testing in PBS resulted in a significant reduction in the mean fatigue life with increasing stress. The observed trend for the effect of stress and aqueous environment on the fatigue of silicate 13-93 bioactive glass scaffolds is compatible with the stress corrosion crack growth mechanisms for bioactive and conventional silicate glasses[39] which is generally attributed to the stress corrosion of Si–O–Si bonds at the crack tip.[40] In normal walking, the compressive load on the human femur is estimated to be smaller than 4 times the body weight.[41] Assuming a uniform load distribution, a femoral bone cross-sectional area[42] of ~4 cm^2, and a body weight of 70 kg, the stress on an implant in a segmental femoral defect is <8 MPa. The results of the present study indicate that these bioactive glass scaffolds created by robocasting have excellent fatigue resistance at stresses higher than normal physiological stresses.

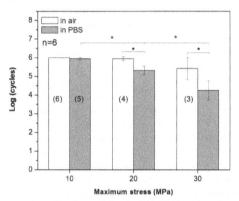

Fig. 6. Fatigue life (average number of cycles to failure) of 13-93 bioactive glass scaffolds tested in air and in phosphate-buffered saline (PBS) under cyclic compressive stresses. The stresses shown are the maximum applied stress in the cyclic loading. (*significant difference between groups, $p < 0.05$). The number in each bar gives the number of samples that survived 10^6 cycles when the test was terminated.

BONE REGENERATION IN RAT CALVARIAL DEFECTS IMPLANTED WITH STRONG POROUS BIOACTIVE GLASS (13-93) SCAFFOLDS

The strong porous bioactive glass (13-93) scaffolds with a grid-like microstructure (**Fig. 4**) were implanted in rat calvarial defects (4.6 mm in diameter) to evaluate their capacity to support bone regeneration. Six weeks postimplantation, the amounts of new bone and mineralization in the defects were evaluated using histomorphometric techniques. The procedures used in the implantation, histologic processing, and histomorphometric analysis are described in detail elsewhere.[43] Three groups of scaffolds were implanted:
(1) As fabricated scaffolds;
(2) Scaffolds pretreated for 3 days in an aqueous phosphate solution (0.25 M K_2HPO_4) to convert a surface layer of the glass (~5 μm) to HA prior to implantation;
(3) Scaffolds pretreated in the phosphate solution (3 days) and loaded with bone morphogenetic protein-2 (BMP2) (1 μg/defect) prior to implantation.
The unfilled defects served as the negative control.

Hematoxylin and eosin (H&E) and von Kossa stained sections of rat calvarial defects implanted with the as-fabricated bioactive glass scaffolds, the scaffolds pretreated in the K_2HPO_4 solution for 3 days, and the pretreated scaffolds loaded with BMP2 are shown in **Fig. 7**. The von Kossa positive area (dark stain) within the defect showed the presence of mineralized bone as well as HA (or calcium phosphate material) resulting from the pretreatment of the scaffolds or conversion in vivo. Because of the limited amount of HA formed in the pretreatment process and the slow conversion of 13-93 bioactive glass in vivo, the von Kossa positive area corresponded generally to the H&E stained areas.

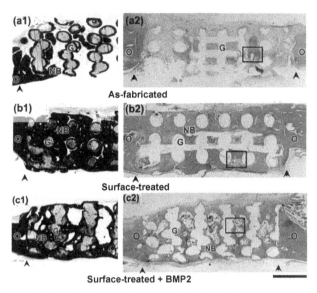

Fig. 7. Von Kossa stained sections (a1−c1) and H&E stained sections (a2−c2) of rat calvarial defects implanted for 6 weeks with three groups of 13-93 bioactive glass scaffolds. (a1, a2): as-fabricated scaffolds; (b1, b2) scaffolds pretreated 3 days in K_2HPO_4 solution; (c1, c2) pretreated 3 days in K_2HPO_4 solution and loaded with BMP-2 (1 μg/defect). Scale bar =1 mm. G = bioactive glass; NB = new bone; O = old (host) bone; arrowheads indicate the edges of old bone.

All implants showed the formation of new bone into the edges (periphery) of the implants (adjacent to the old bone), indicating good integration of the implants with the surrounding bone. New bone was observed mainly within the interior pores of the implants composed of the as-fabricated scaffolds (**Fig. 7a1, a2**). In comparison, the implants composed of the pretreated scaffolds and the pretreated scaffolds loaded with BMP2 showed a better capacity to support new bone formation (**Fig. 7b1−c2**). Blood vessels were observed within all of the implanted scaffolds in the defects (results not shown). When normalized to the total pore area in the section, the amount of new bone formed in the as-fabricated scaffolds after the six-week implantation was 32 ± 13% (**Table III**). In comparison, the percent new bone formed in the scaffolds pretreated in the K_2HPO_4 solution for 3 days was 57 ± 14%, while the amount of new bone in formed in the scaffolds pretreated for 3 days and loaded with BMP-2 was 61 ± 8%. The amount of new bone formed in the defects implanted with the pretreated scaffolds with or without BMP2 loading was significantly higher than that in the as-fabricated scaffolds ($p < 0.05$).

The effectiveness of the three-day pretreatment in enhancing bone regeneration might be related to the composition of the converted surface layer of the scaffold. Upon implantation, the HA surface of the pretreated scaffold can interact immediately with biomolecules and cells in the vicinity of the defect. In comparison, the as-fabricated scaffold has a silicate glass surface that converts first to a silica-rich surface, then to an amorphous calcium phosphate layer, and finally to HA. This difference in surface composition could delay the initial response of the as-fabricated scaffold to biomolecules and cells when compared to pretreated scaffold, and presumably influence the capacity of the scaffold to regenerate bone.

Table III. Amount of new bone formed in rat calvarial defect implanted with the three groups of bioactive glass (13-93) scaffolds at 6 weeks, expressed as a percent of the available pore area in the scaffolds and the total defect area.

Scaffold group	New bone	
	% of available pore area	% of total defect area
As fabricated	32 ± 13	18 ± 8
Pretreated in phosphate solution	57 ± 14	33 ± 10
Pretreated in phosphate solution + BMP2	61 ± 8	35 ± 3

The as-fabricated scaffolds with the grid-like microstructure used in this study showed a better capacity to support new bone formation when compared to 13-93 scaffolds with a fibrous, trabecular, or oriented microstructure used in previous studies.[26,44] The amount of new bone formed at 6 weeks in the as-fabricated scaffolds with the grid-like microstructure (32%) was similar or greater than in the scaffolds with the fibrous, trabecular, or oriented microstructure which were implanted for twice as long (12 weeks) (17, 25, and 37%, respectively). Pretreating the as-fabricated scaffolds to convert a thin surface layer of the glass to HA or loading the pretreated scaffolds with BMP-2 further enhanced the capacity of the grid-like scaffolds to support bone regeneration.

Based on their promising mechanical properties and their capacity to regenerate bone in osseous defects, the 13-93 bioactive glass scaffolds created with a grid-like microstructure by robocasting in this study are promising in structural bone repair. Experiments are in progress to re-design the architecture of the architecture of the scaffolds to optimize their mechanical properties and to evaluate the capacity of the scaffolds to repair loaded bone defects in suitable animal models in vivo.

SUMMARY AND CONCLUSIONS

Our recent studies have shown the ability to create strong porous scaffolds of silicate bioactive glass using robotic deposition techniques such as freeze extrusion fabrication and robocasting. Scaffolds of 13-93 bioactive glass with a uniform grid-like microstructure created by robocasting have a compressive strength and elastic modulus comparable to human cortical bone, good mechanical reliability in compression (Weibull modulus = 12), and excellent fatigue resistance (fatigue life $>10^6$ cycles under compressive loads greater than normal physiological loads). While the flexural modulus is comparable to that of cortical bone, the flexural strength is much lower. The strong porous scaffolds of 13-93 glass created by robocasting have shown the capacity to support bone regeneration in rat calvarial defects. Six weeks postimplantation, 32% of the pore space in the as-fabricated scaffolds was infiltrated with new bone. Treating the as-fabricated scaffold to convert a thin surface layer (~5 μm) of the glass to hydroxyapatite prior to implantation resulted in a significant increase in new bone formation to 57%. Loading the surface-treated scaffolds with BMP2 also significantly enhanced new bone formation (to 61%). Bioactive glass (13-93) scaffolds created by robocasting, surface-treated in an aqueous phosphate solution or loaded with BMP2, have potential in structural bone repair.

ACKNOWLEDGEMENTS: This work was supported by the National Institutes of Health, National Institute of Arthritis and Musculoskeletal and Skin Diseases (NIH/NIAMS) Grant No. 1R15AR056119.

REFERENCES
[1] P. V. Giannoudis, H. Dinopoulos, and E. Tsiridis, "Bone Substitutes: An Update," *Injury*, **36S**, S20−37 (2005).
[2] C. Laurencin, Y. Khan, and S. F. El-Amin, "Bone Graft Substitutes," Expert Rev Med Devices **3**,

49–57 (2006).

[3]S. F. Hulbert, F. A. Young, R. S. Mathews, J. J. Klawitter, C. D. Talbert, and F. H. Stelling, "Potential of Ceramic Materials as Permanently Implantable Skeletal Prostheses," *J. Biomed. Mater. Res.*, **4**, 433–56 (1970.

[4]V. Karageorgiou and D. Kaplan, "Potential of 3D Biomaterial Scaffolds and Osteogenesis," *Biomaterials*, **26**, 5474–91 (2005).

[5]L. L. Hench, "Bioceramics," *J. Am. Ceram. Soc.*, **81**, 1705–28 (1998).

[6]M. N. Rahaman, D. E. Day, B. S. Bal, Q. Fu Q, S. B. Jung, L. F. Bonewald, and A. P. Tomsia, "Bioactive Glass in Tissue Engineering," *Acta Biomater.*, **7**, 2355–73 (2011).

[7]J. R. Jones, "Review of Bioactive Glass – From Hench to Hybrids," *Acta Biomater.*, **9**, 4457–86 (2012).

[8]A. Hoppe, N. S. Güldal, and Boccaccini, "A Review of the Biological Response to Ionic Dissolution Products from Bioactive Glasses and Glass-Ceramics," *Biomaterials*, **32**, 2757–74 (2011).

[9]P. Habibovic and J. E. Barralet, "Bioinorganics and Biomaterials: Bone Repair," *Acta Biomater.*, **7**, 3013–26 (2011).

[10]J. Will, L. C. Gerhardt, and Boccaccini, "Bioactive Glass-Based Scaffolds for Bone Tissue Engineering," *Adv. Biochem. Eng. Biotechnol.*, **126**,195–226 (2012).

[11]M. N. Rahaman, D. E. Day, R. F. Brown, Q. Fu, and S. B. Jung, "Nanostructured Bioactive Glass Scaffolds for Bone Repair," *Cer. Eng. Sci. Proc.*, **29** (7), 211–25 (2008).

[12]Q. Z. Chen, I. D. Thompson, and A. R. Boccaccini, "45S5 Bioglass®-Derived Glass–Ceramic Scaffolds for Bone Tissue Engineering," *Biomaterials*, **27**, 2414–25 (2006).

[13]Q. Fu, M. N. Rahaman, R. F. Brown, B. S. Bla, and D. E. Day, "Mechanical and In Vitro Performance of 13-93 Bioactive Glass Scaffolds Prepared by a Polymer Foam Replication Technique," *Acta Biomater.*, **4**, 1854–64 (2008).

[14]Q. Fu, M. N. Rahaman, B. S. Bal, and R. F. Brown, "Preparation and In Vitro Evaluation of Bioactive Glass (13-93) Scaffolds with Oriented Microstructures for Repair and Regeneration of Load-Bearing Bones," *J. Biomed. Mater. Res. A*, **93**, 1380–90 (2010).

[15]X. Liu, M. N. Rahaman, Q. Fu, and A. P. Tomsia, "Porous and Strong Bioactive Glass (13-93) Scaffolds Prepared by Unidirectional Freezing of Camphene-based Suspensions," *Acta Biomater.*, **8**, 415-23 (2012).

[16]T. S. Huang, N. D. Doiphode, M. N. Rahaman, M. C. Leu, B. S. Bal, and D. E. Day, "Porous and Strong Bioactive Glass (13-93) Scaffolds Prepared by Freeze Extrusion Fabrication," *Mater. Sci. Eng. C*, **31**, 1482–9 (2011).

[17]Q. Fu, E. Saiz, and A. P. Tomsia, "Direct Ink Writing of Highly Porous and Strong Glass Scaffolds for Load-bearing Bone Defects Repair and Regeneration," *Acta Biomater.*, **7**, 3547–54 (2011).

[18]A. M. Deliormanlı and M. N. Rahaman, "Direct-write Assembly of Silicate and Borate Bioactive Glass Scaffolds for Bone Repair," *J. Eur. Ceram. Soc.*, 32, 3637–46 (2012).

[19]T. Kokubo, H. Kushitani, S. Sakka, T. Kitsugi, and T. Yamamuro, "Solutions Able to Reproduce In Vivo Surface-structure Changes in Bioactive Glass–Ceramic A-W," *J. Biomed. Mater. Res.*, **24**, 721–34 (1990).

[20]W. Huang, D. E. Day, K. Kittiratanapiboon, and M. N. Rahaman, "Kinetics and Mechanisms of the Conversion of Silicate (45S5), Borate, and Borosilicate Glasses to Hydroxyapatite in Dilute Phosphate Solutions," *J. Mater. Sci. Mater. Med.*, **17**, 583–96 (2006).

[21]A. Yao, D. P. Wang, W. Huang, Q. Fu, M. N. Rahaman, and D. E. Day, "In Vitro Bioactive Characteristics of Borate-Based Glasses with Controllable Degradation Behavior," *J. Am. Ceram. Soc.*, **90**, 303–6 (2007).

[22]R. F. Brown, D. E. Day, T. E. Day, S. B. Jung, M. N. Rahaman, and Q. Fu, "Growth and Differentiation of Osteoblastic Cells on 13-93 Bioactive Glass Fibers and Scaffolds," *Acta Biomater.*, **4**, 387–96 (2008).

[23]R. F. Brown, M. N. Rahaman, A. B. Dwilewicz, W. Huang, D. E. Day, Y. Li, and B. S. Bal, "Conversion of Borate Glass to Hydroxyapatite and its Effect on Proliferation of MC3T3-E1 Cells," *J. Biomed. Mater. Res. A*, **88**, 392–400 (2009).

[24]Q. Fu, E. Saiz, M. N. Rahaman, and A. P. Tomsia, "Bioactive Glass Scaffolds for Bone Tissue Engineering: State of the Art and Future Perspectives," *Mater. Sci. Eng. C*, **31**, 1245–56 (2011).

[25]Q. Fu, M. N. Rahaman, H. Fu, and X. Liu, "Silicate, Borosilicate, and Borate Bioactive Glass Scaffolds with Controllable Degradation Rates for Bone Tissue Engineering Applications, I: Preparation and In Vitro Degradation," *J. Biomed. Mater. Res. A*, **95**, 164–71 (2010).

[26]L. Bi, S. B. Jung, D. E. Day, K. Neidig, V. Dusevich, J. D. Eick, and L. F. Bonewald, "Evaluation of Bone Regeneration, Angiogenesis, and Hydroxyapatite Conversion in Critical-Sized Rat Calvarial Defects Implanted with Bioactive Glass Scaffolds," *J. Biomed. Mater. Res.* A, **100**, 3267–75 (2012).

[27]L. Bi, M. N. Rahaman, D. E. Day, Z. Brown, C. Samujh, X. Liu, A. Mohammadkhah, V. Dusevich, J. D. Eick, and L. F. Bonewald, "Effect of Borate Bioactive Glass Microstructure on Bone Regeneration, Angiogenesis, and Hydroxyapatite Conversion in a Rat Calvarial Defect Model," *Acta Biomater.*, (2013); in press.

[28]X. Liu, M. N. Rahaman, G. E. Hilmas, B. S. Bal, "Mechanical Properties of Bioactive Glass Scaffolds Fabricated by Robotic Deposition for Structural Bone Repair," *Acta Biomater.*, 2013; http://dx.doi.org/10.1016/j.actbio.2013.02.026.

[29]K. U. Lewandrowski, D. L. Wise, M. J. Yaszemski, J. D. Gresser, D. J. Trantolo, and D. E. Altobelli, *Tissue Engineering and Biodegradable Equivalents: Scientific and Clinical Applications*. New York, Marcel Dekker, 2002.

[30]F. J. Martínez-Vázquez, F. H. Perera, P. Miranda, A. Pajares, and F. Guiberteau, "Improving the Compressive Strength of Bioceramic Robocast Scaffolds by Polymer Infiltration," *Acta Biomater.*, **6**, 4361–8 (2010).

[31]P. Miranda, A. Pajares, E. Saiz, A. P. Tomsia, and F. Guiberteau, "Mechanical Properties of Calcium Phosphate Scaffolds Fabricated by Robocasting," *J. Biomed. Mater. Res. A*, 85, 218–27 (2008).

[32]Y. Shanjani, Y. Hu, R. M. Pilliar, and E. Toyserkani, "Mechanical Characteristics of Solid-Freeform-Fabricated Porous Calcium Polyphosphate Structures with Oriented Stacked Layers," *Acta Biomater.*, 7, 1788–96 (2011).

[33]N. Verdonschot and R. Huiskes, "Creep Behavior of Hand-Mixed Simplex P Bone Cement Under Cyclic Tensile Loading," *J. Appl. Biomater.*, **5**, 235–43 (1994).

[34]R. D. Crowninshield, R. A. Brand, R. C. Johnston, and J. C. Milroy, "The Effect of Femoral Stem Cross-Sectional Geometry on Cement Stresses in Total Hip Reconstruction," *Clin. Orthop. Rel. Res.*, **146**, 71–7 (1980).

[35]I. Quintana-Alonso, S. P. Mai, N. A. Fleck, D. C. H. Oakes, and M. V. Twigg, "The Fracture Toughness of a Cordierite Square Lattice," *Acta Mater.*, **58**, 201–7 (2010).

[36]W. Bonfield, "Elasticity and Viscoelasticity of Cortical Bone," In: G. W, Hastings and P. Ducheyne editors, *Natural and Living Biomaterials*. Boca Raton, FL, CRC Press, 1985.

[37]Y. Zhang, H. Xu, S. Takagi, and L. Chow, "In-situ Hardening Hydroxyapatite-based Scaffold for Bone Repair," *J. Mater. Sci. Mater. Med.*, **17**, 437–45 (2006).

[38]F. Pernot, P. Etienne, F. Boschet, and L. Datas, "Weibull Parameters and the Tensile STrength of Porous Phosphate Glass–Ceramics," *J. Am. Ceram. Soc.*, **82**, 641–8 (1999).

[39]D. R. Bloyer, J. M. McNaney, R. M. Cannon, E. Saiz, A. P. Tomsia, and R. O. Ritchie, "Stress–Corrosion Crack Growth of Si–Na–K–Mg–Ca–P–O Bioactive Glasses in Simulated Human Physiological Environment," *Biomaterials*, **28**, 4901–11 (2007).

[40]T. A. Michalske and S. W. Freiman, "A Molecular Mechanism for Stress Corrosion in Vitreous Silica," *J. Am. Ceram. Soc.*, **66**, 284–8 (1983).

[41]G. N. Duda, E. Schneider, and E. Y. S. Chao, "Internal Forces and Moments in the Femur During Walking," *J. Biomech.*, **30**, 933–41 (1997).

[42]J. Rittweger, G. Beller, J. Ehrig, C. Jung, U. Koch, J. Ramolla, F. Schmidt, D. Newitt, S. Majumdar, H. Schiessl, and D. Felsenberg, "Bone–Muscle Strength Indices for the Human Lower Leg," *Bone*, **27**, 319–26 (2000).

[43]X. Liu, M. N. Rahaman, Y. Liu, B. S. Bal, and L. F. Bonewald, "Enhanced Bone Regeneration in Surface-modified and BMP-loaded Bioactive Glass (13-93) Scaffolds in a Rat Calvarial Defect Model," *Acta Biomater.*, 2013; DOI: 10.1016/j.actbio.2013.03.039.

[44]X. Liu, M. N. Rahaman, and Q. Fu, "Bone Regeneration, Mineralization, and Mechanical Response of Bioactive Glass (13-93) Scaffolds with Oriented and Trabecular Microstructures Implanted in rat Calvarial Defects," *Acta Biomater.*, **9**, 4889–98 (2013).

FABRICATION, CHARACTERIZATION AND *IN-VITRO* EVALUATION OF APATITE-BASED MICROBEADS FOR BONE IMPLANT SCIENCE

J. Feng[1], M. Chong[2], J. Chan[3,4], Z.Y. Zhang[5], S.H. Teoh[2], E.S. Thian[1,*]

[1]Department of Mechanical Engineering
National University of Singapore
Singapore 117 576, Singapore

[2] Division of Bioengineering
School of Chemical and Biomedical Engineering
Nanyang Technological University
Singapore 637 459, Singapore

[3]Department of Obstetrics and Gynaecology
National University of Singapore
Singapore 119 228, Singapore

[4] Department of Reproductive Medicine
KK Women's and Children's Hospital
Singapore 229 899, Singapore

[5] Department of Plastic and Reconstructive Surgery
Shanghai 9th People's Hospital
Shanghai Jiao Tong University
Shanghai 200011, China

*Corresponding Author - Email: mpetes@nus.edu.sg

ABSTRACT
The fabrication and characterization of apatite microbeads, which are intended to be used for bone tissue engineering, will be featured in this work. Hydroxyapatite-Alginate (HA-Alg) suspension was extruded dropwise into a calcium chloride ($CaCl_2$) crosslinking solution. The HA-Alg beads were then sintered using a 4-stage sintering profile to form porous microbeads of uniform size. Different concentrations of Alg solution, $CaCl_2$ crosslinking solution, and HA-to-Alg ratios, were used. Results showed that Alg concentration affected the ability for particles to form spherical microbeads during the drying process. HA-to-Alg ratio affected the packing density of the microbeads, and sufficient HA-to-Alg content would yield more spherical microbeads. $CaCl_2$ concentration affected the extent of crosslinking within the HA-Alg microbeads. Insufficient Ca^{2+} ions available would cause microbeads having deep furrows, indicating incomplete crosslinking in the HA-Alg beads. Thermal analyses were conducted to characterize the HA-Alg microbeads and subsequently, a multi-stage sintering profile was designed to yield porous microbeads with less cracks and better necking. The sintered microbeads measured in the range of $850 - 1120$ μm. X-ray diffraction (XRD) results showed that HA microbeads remained phase-pure even after sintering process since no other secondary phases of calcium phosphate were detected. It also revealed that HA crystallinity increased with sintering temperature. Fourier transform infrared (FTIR) analysis indicated several sharp phosphate bands coupled with a hydroxyl band (all belonging to HA), and several carbonate

bands. A preliminary *in-vitro* test revealed that human fetal mesenchymal stem cells (hfMSCs) grew well and remained viable with culture time.

Keywords: alginate; hydroxyapatite; microbeads; porous; stem cells

1. INTRODUCTION

Mesenchymal stem cells (MSCs) are multipotent cells, which are readily isolated and expanded from bone marrow, with a well-defined osteogenic differentiation pathway[1,2], and demonstrated great potential for tissue engineering applications[3]. Existing expansion technique based on culture flasks requires trypsin-mediated multiple passages of monolayer cells to achieve sufficient cell numbers. This method is costly, time-consuming, and susceptible to contamination due to numerous passages needed to generate sufficient cells for transplantation. Furthermore, the removal of extracellular matrix laid down during stem cell expansion by repeated trypsinisation is likely to lead to reduced intra-cellular signalling responsible for cell viability, proliferation and differentiation[4]. Thus, microbeads that provide a high surface area for cell attachment has generated great interest nowadays[5-7].

The currently available microbeads are made of either plastic polymers or glass-based matrices[8-10], mainly designed and developed for *in-vitro* applications in pharmaceutical industries, and may not be suitable for therapeutic implantation. Certainly, the need for such technology is potentially huge. Recently, there has been some works reporting on the fabrication of microbeads using HA as the material[11,12].

Bioceramics have been widely used in the medical applications. The use of calcium phosphate ceramics in bone regeneration, either alone or in combination with a polymeric phase, has become a common practice since these materials generally provide good biological responses and adequate mechanical properties. Clinically, bioceramics such as hydroxyapatite (HA), tricalcium phosphate (TCP), and bioglass have been used as bone implants[13]. HA, for its well-known osteoconductive property, has been chosen as an ideal biomaterial that can promote bone growth and repair. However, the use of this bioactive ceramic for defect filler is mainly in powder and/or block forms. Although powder ceramics remain as the preferred choice for filling small irregular defects, the therapeutic effect of the filling implant is lost via migration of particles from the defect site[14].

Alginates (Alg) are natural occurring polysaccharides derived from brown algae. They contain the copolymers of β-D-mannuronic acid (M unit) and α-L-guluronic acid (G unit) arranged in different proportions and sequences, which will determine the physical properties of the alginate molecules[15]. Alg are also capable of forming relatively stable hydrogels through ionotropic gelation. This crosslinking process can be carried out under very mild conditions, at low temperatures, and in the absence of organic solvents, and hydrogels of different shapes can be prepared. During crosslinking, divalent cations such as Ca^{2+} ions co-operatively bind between the G-blocks of adjacent Alg chains, creating ionic inter-chain bridges, which cause gelling of aqueous Alg solution[16]. Alg present a viable biocompatible polymer candidate to encapsulate HA into nanosized spherical particles. Several therapeutic agents including antibiotics, enzymes and growth factors, have already been successfully incorporated in Alg gels, retaining a high percentage of biological activity[17].

In this work, the authors use a suspension of HA-Alg, which is extruded dropwise into a $CaCl_2$ crosslinking solution. The subsequent HA-Alg microbeads are washed and dried, followed by a sintering process to burnt-off the polymer. The purpose of this work is to conduct morphological and chemical characterization of the sintered HA microbeads, and identify optimal parameters which can produced good microbeads intended for bone tissue engineering.

2. MATERIALS AND METHODS

2.1. Synthesis of phase-pure HA

HA was synthesized via a wet precipitation route. Calcium hydroxide [$Ca(OH)_2$] and orthophosphoric acid (H_3PO_4) were weighed and dissolved in deionised (DI) water to form solutions. 1 l of an aqueous H_3PO_4 (0.6 M) was added dropwise into 1 l of an aqueous $Ca(OH)_2$ (1 M) under continuous stirring at room temperature. Concentrated ammonia solution was added to maintain a pH value above 10.5. The solution was left to age for 2 weeks before subjecting it to an autoclave process for 2 h. The resulting precipitate was then dried in an oven at 70 °C for 12 h.

2.2. Synthesis of HA microbeads

Varying proportions of Na-Alg and HA powders were used (Table I), and the resulting solution was extruded dropwise into a $CaCl_2$ crosslinking solution of varying concentrations using a droplet extrusion device (internal diameter of nozzle = 0.89 mm). Na-Alg was dissolved in DI water for 30 mins using a magnetic stirrer until total homogenization was achieved. HA powder was then added to the Na-Alg solution, and stirred at 50 °C for 8 h using a hot plate stirrer, to ensure thorough homogenization. The solution containing HA and Na-Alg was extruded dropwise using a roller pump into $CaCl_2$ crosslinking solution, whereby spherical microbeads were formed instantaneously.

The solution containing HA-Alg microbeads and $CaCl_2$ was then allowed to mix for 2 h under gentle magnetic stirring to ensure crosslinking completion. HA-Alg microbeads were retrieved from the solution and washed twice with distilled water to remove any $CaCl_2$. These microbeads were left to dry overnight at room temperature before subjecting to a multi-stage sintering process.

A multi-stage sintering profile (Figure 1) was designed so as to produce microbeads with better characteristics. The stages were designed after the melting temperature of Alg were obtained so as to allow for a more gradual burnt-off as well as to allow for longer time for the necking process between HA particles to occur.

Table I. HA microbeads fabricated using different Alg concentrations, HA contents and $CaCl_2$ concentrations

Sample	Alg Concentration (g/ml)	HA Content (wt.%)	$CaCl_2$ Concentration (mol/dm^3)
1	0.0128	20	0.1
2	0.015	20	0.1
3	0.03	20	0.1
4	0.03	40	0.1
5	0.03	40	0.5
6	0.03	50	0.1
7	0.03	50	0.5
8	0.03	40	0.5

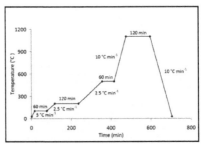

Figure 1. 4-stage sintering profile for HA-Alg microbeads.

2.3. Characterisation of HA microbeads

2.3.1. Optical microscopy analysis
The size of the microbeads was estimated using an optical microscope with a calibrated built-in measurement scale. The mean diameter was determined from a sample of twenty microbeads.

2.3.2. Scanning electron microscopy analysis
The microbead's morphology was analysed using a field emission scanning electron microscope (FE-SEM, Hitachi S-4300), operating at an accelerating voltage of 15 kV. The samples were sputtered-coated with a thin layer of gold before examination.

2.3.3. Energy dispersive X-ray spectroscopy analysis
Preliminary surface characterization of the HA microbeads was analysed using an energy dispersive X-ray spectroscopy (EDS). Samples were loaded into the SEM machine operating at an accelerating voltage of 15 kV, and a working distance of 10 mm. The elemental quantitative analysis used an automatic background subtraction, and a ZAF correction matrix has been used to calculate the elemental composition in weight percentage.

2.3.4. Thermogravimetric analysis
Thermogravimetric analysis (TGA) was conducted for the HA-Alg microbeads to determine its thermal stability and burnt-off characteristics. Sample weight was 14 mg, and a platinum pan was used. TGA was conducted between 25 and 1200 °C at a heating rate of 10 °C/min, in an argon atmosphere.

2.3.5. Differential scanning calorimetry analysis
Differential thermal analysis (DTA) analysis was conducted to determine the melting temperature of HA-Alg microbeads. 22 mg of HA-Alg microbeads was added onto an aluminium pan, and heated from 25 to 500 °C at a rate of 10 °C/min.

2.3.6. X-ray diffraction analysis
The phase composition of HA microbeads was investigated using an X-ray diffractometer (XRD, Shimadzu X-ray diffractometer, Model 6000). The microbeads were crushed and compacted before loading onto the machine. CuK_α radiation ($\lambda = 1.5406$ nm) at a scanning rate of 0.3 °/min was used over a 2θ range of $20 - 40$ °, with a sampling interval of 0.05 °, operating

at 30 mA and 40 kV. Phases were identified by comparison of the experimental data with the reference data from the International Centre for Diffraction Data.

2.3.7. Fourier transform infrared spectroscopy analysis

Chemical characterisation was performed using a fourier transform infrared spectroscometer (FTIR, Varian 3100 spectrometer). For this purpose, HA microbeads were crushed and analysed using potassium bromide pellets. A wavelength ranging from 400 to 4000 cm^{-1} with a spectral resolution of 16 cm^{-1} was used.

2.4. Cell viability on HA microbeads

Cell viability behaviour on HA microbeads was evaluated using human fetal mesenchymal stem cells (hfMSCs). 18 mg of HA microbeads and 4 mg of Cytodex 3 microbeads (GE Healthcare, USA) were added to each well. The calculation of weights to be used was based on the total surface area added per well. For HA microbeads, this was 0.55 cm^2/mg, and 2.7 cm^2/mg for Cytodex 3. This allowed for the total surface area per well of each microbead type to be 10 cm^2. 1.0×10^5 cells were then added to each well, such that the seeding density for both microbead type was 1.0×10^4 cells/cm^2. hfMSCs were cultured on sterile microbeads in Dulbecco's modified Eagle's medium (DMEM, Sigma, USA)-GlutaMAX (GIBCO, USA) supplemented with 10% fetal bovine serum (FBS), 50 U/ml penicillin, and 50 mg/ml streptomycin (GIBCO, USA) (referred as D10 medium) and incubating at 37 °C in a humidified atmosphere of 95 % air and 5 % carbon dioxide for 14 days, with medium change every 3 – 4 days.

Cell viability was assessed quantitatively using PrestoBlue assay (Invitrogen, USA), which measured cell viability through the reduction of resazurin to resorufin. On the designated time points, 10 % PrestoBlue reagent was added to each well and incubated for 25 min at 37 °C. Each time point was measured in triplicates. Absorbance at 570 nm, referenced at 600 nm was read using a microplate reader (Tecan, USA). The intensity cross-referenced to a standard calibration curve of hfMSC count against absorbance done at the beginning of the study to obtain the live cell count at each time point. The qualitative analysis of cell viability on HA microbeads was performed by fluorescein di-acetate/propidium iodide (FDA/PI) staining. Briefly, 500 μl of HA microbeads was extracted from a 24-well plate, washed twice with phosphate buffer saline solution, and stained with 8 mg/ml FDA. The samples were then incubated for 15 min, washed and stained with 2 mg/ml PI, and incubated for another 2 min. The samples were given a final wash before examining under a confocal laser scanning microscope (CLSM, Olympus FV1000, Japan). FDA stained viable cell cytoplasm green, and PI stained necrotic and apoptotic cell nuclei red.

3. RESULTS

3.1. Pre-sintered HA-Alg microbeads

The first step to HA microbead fabrication involved HA-Alg bead formation. The following results report on the physical morphology of HA-Alg beads prepared under various settings.

3.1.1. Low HA content, low Alg concentration

The initial batch of HA-Alg beads were fabricated using 20 wt.% HA and 0.015 g/ml Alg. Flow through the tubing during droplet extrusion was rapid as the solution exhibited low viscosity. The fabricated beads in the 0.1 M CaCl₂ solution appeared to have good spheroidicity, and they remained intact after removing from the solution and washed. However, upon drying, HA-Alg beads appeared to be more disc-like, with flattened structure. This could be attributed to excessive water loss from the beads during drying, and also insufficient Alg content to maintain its integrity of the bead. The appearance of white specks within most HA-Alg beads indicated insufficient homogenisation of HA within the Alg matrix. The size of the pre-sintered particles ranged from $1380 - 2270$ μm, with a mean of 1.8 mm (n = 20). They displayed high tendency to clump together and form agglomerates during drying.

3.1.2. High HA content, high Alg concentration
After the initial batch of fabrication, HA-Alg beads were fabricated using 40 and 50 wt.% HA. A higher concentration of Alg solution (0.03 g/ml) was also used. Beads that were formed during the crosslinking process maintained their shape after drying, and were more spherical compared to beads prepared using low Alg concentration. In addition, the beads did not tend to aggregate. The variation of HA content between 40 and 50 wt.% did not yield any significant change in size. The size of the pre-sintered beads ranged from $980 - 1620$ μm, with a mean of 1.4 mm (n = 20). This value was slightly smaller than beads formed using low Alg concentration, which could be attributed to better crosslinking within the Alg polymer chains.

Some HA-Alg beads were crushed and examined using FE-SEM. The internal morphology of the beads revealed specks of HA embedded within the Alg matrix (Figure 2). EDS analysis confirmed that the grey areas consisted mainly of carbon and oxygen whilst the brighter specks contained calcium and phosphorus. The HA distribution within the matrix was homogeneous, with no obvious aggregation of HA particles being observed in any part of the bead.

A higher concentration of Alg generally led to better microbead formation, while a higher HA content led to better distribution of HA within the microbead. However, as Alg concentration and HA content increased, viscosity of the HA-Alg is increased, and this would lead to an increase in the incidence of tube occlusions. The authors have found that 0.03 g/ml Alg and 40 wt. % HA represented the optimal parameters for obtaining spherical sintered HA microbeads with minimal risks of tube occlusions.

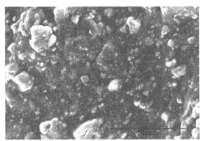

Figure 2. SEM image of crushed pre-sintered HA-Alg microbead (50 wt.% HA, 0.03 g/ml Alg) to show internal morphology.

3.1.3 Effect of $CaCl_2$ concentration
To investigate the effect of $CaCl_2$ concentration on bead formation, two batches of HA-Alg suspension (40 wt.% HA, 0.03 g/ml Alg) were extruded into crosslinking solution of 0.1 and 0.5 M $CaCl_2$, respectively.

HA-Alg beads prepared using 0.1 M $CaCl_2$ demonstrated pits and furrows (Figure 3a). When a higher concentration of 0.5 M $CaCl_2$ was used, none of the beads exhibited pits (Figure 3b). This phenomenon suggested that using a low concentration of $CaCl_2$ could result in incomplete crosslinking of the beads. Beads prepared using 0.1 or 0.5 M CaCl$_2$ had an average size of 1.4 mm.

The higher concentration of CaCl$_2$ is suggested to lead to the stronger crosslinking produced during ionotropic gelation thereby causing denser HA particle-particle packing. This would be beneficial towards maintaining structural integrity of the sintered HA microbeads as the necking regions within the sintered microbeads would be stronger.

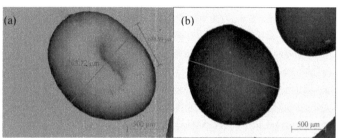

Figure 3. Pre-sintered HA-Alg microbeads fabricated using (a) 0.1 M and (b) 0.5 M CaCl$_2$ crosslinking solution.

3.2. Thermal analysis

To investigate the thermal stability of HA-Alg beads, TGA (Figure 4a) and DTA (Figure 4b) were conducted for beads containing 40 wt.% HA and 0.03 g/ml Alg. The HA-Alg beads lost 60 % by weight in total during the sintering process from 23 to 1200 °C. Weight loss occurred almost immediately at 50 °C, and continued all the way to about 1000 °C. The first major weight loss occurred at 200 °C, which registered a weight change of approximately 35 %. This corresponded to the rupture of the Alg monomers as well as the removal of water within the Alg beads. A second weight loss occurred approximately 400 °C, which accounted for 20 % weight loss. A third weight transition occurred at 640 °C, which accounted for a 5 % weight loss. The second and third weight losses could be due to the thermal decomposition of sodium and calcium Alg complexes to intermediate compounds and subsequently, breakdown of these intermediates to metal oxides. The suggested mechanisms were as follows[18]:

At temperature ≤ 200 °C,
$$C_{12}H_{14}O_{11}M.xH_2O \rightarrow C_{12}H_{14}O_{11}M + xH2O$$
At temperature > 200 °C,
$$C_{12}H_{14}O_{11}M + 10CO_2 \rightarrow M(COO)_2 + 10CO_2 + 7H_2O$$
At temperature > 350 °C,
$$M(COO)_2 \rightarrow MO \text{ (metal oxide)} + 2CO_2$$

The stages of decomposition were supported using the DTA graph. The initial drop in energy value up to 82 °C was an endothermic process, which could be explained by the absorption of energy by the beads to break the bonds needed for the liberation of water molecules. A small peak observed at around 200 °C corresponded to the stage where the HA-Alg broke down into an intermediate metal-oxalate. Subsequently, the strong exothermic peak at 420 °C corresponded to the third stage where the metal-oxalate began to form the more stable metal oxide. Beyond 1000 °C, no further weight change occurred, and the sample remained stable up to 1200 °C.

By designing a multi-stage sintering profile, sufficient time would be allowed for the different reactions to take place. This would be crucial in obtaining pure HA microbeads with minimal impurities. Furthermore, by allowing sufficient time at the hold temperatures, defects and cracks in the sintered HA microbeads could be minimised.

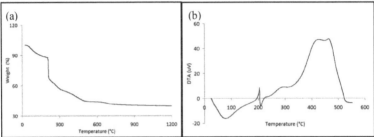

Figure 4. Thermal analyses of HA-Alg microbeads (40 wt.% HA, 0.03 g/ml Alg).
(a) TGA graph, and (b) DTA graph.

3.3. Sintered HA microbeads

HA-Alg microbeads containing 40 wt.% HA and 0.03 g/ml Alg were sintered using a 4-stage sintering profile (Figure 1). SEM image (Figure 5a) revealed that the sintered microbeads displayed good spherical morphology, ranging 850 – 1120 μm with a mean of 1040 μm (n = 20). Some microbeads displayed occasional surface cracks. The cross-sectional view demonstrated that these microbeads displayed high interconnectivity, and HA necking regions were clearly visible (Figure 5b). Distinct micropores could also be observed. Similar to the surface, the inner core displayed highly packed, interconnected HA structure that was rough and porous.

Figure 5. Sintered HA microbeads (40 wt.% HA, 0.03 g/ml Alg).
(a) Normal view, and (b) Cross-sectional view.

3.4. XRD analysis

XRD was done for the autoclaved HA powder, and sintered HA microbeads containing 40 wt.% HA. The results were then compared with the standard XRD patterns of HA (JCPDS 9-432). All the peaks corresponded to that of apatite. The peaks obtained for the sintered microbeads were well-defined and distinct, indicating that HA had a good crystalline phase. By comparing the XRD patterns of the sintered microbeads with the HA powder that was used to synthesise the microbeads, it could be seen that after sintering, the peaks obtained from the microbeads represented a more crystalline phase, suggesting that an increase in temperature increased the amount of crystallinity. Phases corresponding to tricalcium phosphate and tetracalcium phosphate were not observed, confirming that HA in the microbeads remained chemically stable during the sintering process.

Furthermore, a phase-pure apatite material would be advantageous towards promoting osteoconductivity and its chemical similarity to the mineral phase of natural bone would address issues of biocompatibility[20]. Moreover, phase pure apatite has low degradation rate, making it an ideal choice for a long-term tissue engineered solution that would not only encourage bone regeneration, but also undertake part of the defect site's structural functions as the wound site heals. Its high thermal stability also makes it easily autoclavable and eliminates the need for complex sterilization techniques.

3.5 FTIR analysis

FTIR was conducted for the autoclaved HA powder, and sintered microbeads containing 40 wt.% HA. For the HA microbeads, sharp bands at 567, 603 and 632 cm^{-1} were detected and corresponded to the v_4 phosphate bands. These peaks were also present in the autoclaved HA powder, but were less distinct, implying that crystallinity increased with sintering temperature. Furthermore, sharp bands at 960, 1047 and 1093 cm^{-1} were seen for both HA powder and microbeads, which were definitive of the v_3 phosphate band, present in HA. The characteristic bands of the COO- groups at 1318, 1420 and 1619 cm^{-1} as well as the –CCH group at 809, 903 and 945 cm^{-1} were absent, indicating that there was no significant contamination of Alg residues being present[19]. Absence of any residual Alg functional groups would reduce or even eliminate any possible adverse immunogenic/allergic response by the body.

Referring to HA powder, the broad band at 3500 cm^{-1} suggested that there were inclusions of water as well as structural hydroxyl group. Upon sintering, this band resolved into a sharp peak at 3571 cm^{-1}, which could be seen for the sintered microbeads. This peak was the characteristic band for hydroxyl stretching.

Bands present at 875, 1423, 1454 and 1641 cm^{-1} corresponded to the presence of a carbonate phase, implying that carbonate had been incorporated into the HA structure. This effect could be due to the decomposition of Alg whereby carbonate ions substitute for either phosphate groups or hydroxyl group.

3.6 Cell viability

hfMSCs grew well on the HA microbeads with culturing time (Figure 6). Cells remained viable after day 7. By day 14, hfMSCs seeded on HA microbeads were confluent throughout the surface. Prestoblue cell viability assay also showed that cells proliferated well when seeded onto HA microbeads (Figure 7). Compared with Cytodex 3 microbeads, HA microbeads exhibited similar growth profiles, with the exception that log phase growth was obtained on day 5 – 7. On day 9, the number of viable cells on HA microbeads was 1.4-fold higher than that of Cytodex 3 microbeads. Although these observations required in-depth analysis, it had been suggested that surface roughness[21], microporosity[22] and crystallinity[23] of HA were factors that could influence mesenchymal stem cell adhesion, proliferation and differentiation characteristics.

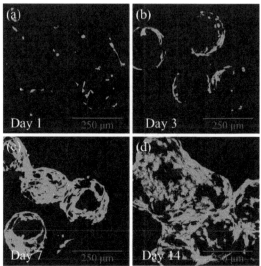

Figure 6. CLSM images demonstrating viable hfMSCs growing on HA microbeads at culture day (a) 1, (b) 3, (c) 7 and (d) 14.

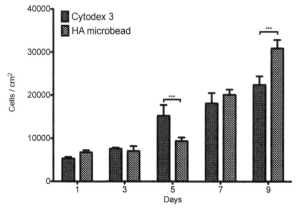

Figure 7. PrestoBlue proliferation assay of hfMSCs cultured on Cytodex 3 and HA microbeads (***p < 0.001).

5. CONCLUSIONS

The fabrication technique for HA microbeads was reported. This involved mixing HA with a solution of Alg. The suspension was then extruded via a droplet-extrusion method into a $CaCl_2$ crosslinking solution. Subsequently, the HA-Alg microbeads were subjected to a sintering profile, to produce the sintered HA microbeads. Various processing parameters such as HA content, Alg concentration and $CaCl_2$ crosslinking concentration would influence the overall morphology and physicochemical characteristics of the microbeads, and these parameters could be controlled to produce the desired HA microbeads. The authors proposed that a 40 wt.% HA, 0.03 g/ml Alg concentration, 0.5 M $CaCl_2$ crosslinking solution, and a 4-stage sintering profile represented the optimal parameters. SEM images revealed that the sintered microbeads displayed a rough surface morphology with presence of interconnected pores. In addition, sintered microbeads displayed high phase-purity as determined by XRD and FTIR analyses. These properties could be ideal for cell adhesion and proliferation. Initial *in-vitro* study involving hfMSCs seeded on HA microbeads was performed. Results suggested better proliferative potential of HA microbeads as compared to Cytodex 3. While further studies have to be conducted to confirm these observations, HA microbeads could offer a viable and highly efficient platform for applications involving direct injectable cell expansion system for bone tissue engineering.

ACKNOWLEDGEMENTS
This work is supported by the Singapore Ministry of Health's National Medical Research Council under its NMRC New Investigator Grant NIG10nov032.

REFERENCES
[1]Frank O, Heim M, Jakob M, Barbero A, Schafer D, Bendik I, Dick W, Heberer M, Martin I. Real-time quantitative RT-PCR analysis of human bone marrow stromal cells during osteogenic differentiation *in vitro*. *J Cell Biochem,* **85,** 737-746 (2002).
[2]Qi H, Aguiar DJ, Williams SM, Pean AL, Pan W, Verfaillie CM. Identification of genes responsible for osteoblast differentiation from human mesenchymal progenitor cells. *Proc Natl Acad Sci USA,* **100,** 3305-3310 (2003).
[3]Mauney JR, Volloch V, Kaplan DL. Role of adult mesenchymal stem cells in bone tissue engineering applications: Current status and future prospects. *Tissue Eng,* **11,** 787-802 (2005).
[4]Reddig PJ, Juliano RL. Clinging to life: Cell to matrix adhesion and cell survival. Cancer Metastasis Rev (2005) 24 (3): 425-439
[5]Abbah SA, Lu WW, Peng SL, Aladin DMK, Li ZY, Tan WK, Cheung KMC, Luk KDK, Zhou GQ. Extracellular matrix stability of primary mammalian chondrocytes and intervertebral disc cells cultured in alginate-based mircobead hydrogels. *Cell Transplantation,* **17,** 1181-1192 (2008).
[6]Liu JY, Hafner J, Dragieva G, Burg G. High yields of autologous living dermal equivalents using porcine gelatine microbeads as microbeads for autologous fibroblasts. *Cell Transplantation* **15,** 445-451 (2006).
[7]Zangi L, Rivkin R, Kassis I, Levdansky L, Marz G, Gorodetsky R. High-yield isolation, expansion and differentiation of rat bone marrow-derived mesenchymal stem cells with fibrin microbeads. *Tissue Eng,* **12,** 2343-2354 (2006).
[8]Griffith B. Scale-up of suspension and anchorage-dependent animal cells. *Molecular Biotechnology,* **17,** 225-238 (2001).
[9]Johanson A, Nielson V. Biosilon a new microcarrier. *Dev Biol Stand,* **46,** 125-129 (1980).
[10]Varani J, Dame M, Beals TF, Wass JA. Growth of three established cell lines on glass microbeads. *Biotech Bioeng,* **25,** 1359-1372 (1983).
[11]Perez RA, Valle SD, Altankov G, Ginebra MP. Porous hydroxyapatite and gelatin/hydroxyapatite microspheres obtained by calcium phosphate cement emulsion. *J Biomed Mater Res,* **97B,** 156-166 (2011).
[12]Qiu QQ, Ducheyne P, Ayyaswamy PS. New bioactive, degradable composite microspheres as tissue engineering substrates. *J Biomed Mater Res,* **52,** 66-76 (2000).

[13]Nandi SK, Roy S, Mukherjee P, Kundu B, De DK, Basu D. Orthopedic applications of bone graft and graft substitutes: A review. *Indian J Med Res,* **132**, 15-30 (2010).
[14]Sikavitsas VI, Temenoff JS, Mikos AG. Biomaterials and bone mechanotransduction. *Biomater* **22**, 2581-2593 (2001).
[15]Gacesa P. Alginates. *Carbohydrate Polymers,* **8**, 161-182 (1988).
[16]Kurt T, Draget CT. Chemical, physical and biological properties of alginates and their biomedical implications. *Food Hydrocolloids,* **25**, 251-256 (2011).
[17]Gombotz WR, Wee SF. Protein release from alginate matrices. *Adv Drug Delivery,* **31**, 267-285 (1998).
[18]Said AA, Hassan RM. Thermal decomposition of some divalent metal alginate gel compounds. *Polymer Degradation and Stability,* **39**, 393-397 (1993).
[19]Rehman I, Bonfield W. Characterisation of hydroxyapatite and carbonated apatite by photo acoustic FTIR spectroscopy. *J Mater Sci: Mater Med,* **8**, 1-4 (1997).
[20]Best SM, Porter AE, Thian ES, Huang J. Bioceramics: Past, present and for the future. *Journal of the European Ceramic Society,* **28**, 1319-1327 (2008).
[21]Deligianni DD, Katsala ND, Koutsoukos PG, Missirlis YF. Effect of surface roughness of hydroxyapatite on human bone marrow cell adhesion, proliferation, differentiation and detachment strength. *Biomaterials,* **22**, 87-96 (2000).
[22]Rouahi M, Gallet O, Champion E, Dentzer J, Hardouin P, Anselme K. Influence of hydroxyapatite microstructure on human bone cell response. *Journal of Biomedical Materials Research Part A,* **78**, 222-35 (2006).
[23]Yang Y, Dennison D, Ong JL. Protein adsorption and osteoblast precursor cell attachment to hydroxyapatite of different crystallinities. *The International journal of oral & maxillofacial implants,* 20, 187 (2005).

A FUNCTIONALIZED NANOPOROUS ALUMINA MEMBRANE ELECTROCHEMICAL
SENSOR FOR DNA DETECTION WITH GOLD NANOPARTICLE AMPLIFICATION

Weiwei Ye and Mo Yang*
Interdisciplinary Division of Biomedical Engineering, the Hong Kong Polytechnic University,
Hong Kong, P.R. China

ABSTRACT
Nucleic acids based detection has become increasingly important for biomedical
diagnostics, environmental monitoring and food screening. A nanoporous alumina membrane
based sensing platform with gold nanoparticle amplification has been developed for direct DNA
detection. The purpose of this study is to develop a functionalized nanoporous alumina
membrane based sensor for direct DNA detection with gold nanoparticle amplification via
impedance spectroscopy. Single stranded DNA was first covalently immobilized on the nanopore
walls via silane-PEG-NHS linker. The residual NHS group was hydrolyzed to form PEG layer
which could minimize the unspecific DNA binding during hybridization process. Using this
nanoporous alumina membrane based device, single strand DNA (ssDNA) immobilization,
double strand DNA (dsDNA) hybridization and gold nanoparticle conjugated dsDNA
hybridization were successfully monitored via impedance spectroscopy. The result showed that
the conjugation of gold nanoparticle with dsDNA could significantly increase the impedance
signal. This portable PDMS device integrated with functionalized nanoporous alumina
membrane with gold nanoparticle amplification has the potential for nucleic acid in field
detection in the application of food screening and environmental monitoring with high
sensitivity.

INTRODUCTION
As the living pace and pressure increase, many diseases, such as cancers, hepatitis, and
nephritis are threatening people's health. Accurate diagnosis of these diseases in the early stage
plays an important role in medical treatment and health recovery. One efficient diagnosis method
is DNA detection, which provides an effective way in detecting the identification of some virus.
For example, HBV-DNA detection result has shown that replication of hepatitis virus could
reduce HBV infection risk[1].
The essential mechanism of DNA sensor is based on the recognition of target DNA
segment by hybridization with its complementary DNA pair immobilized on the recognition
layer of solid surface. The earliest DNA detection method used enzyme-labeled DNA probe to
detect target DNA. The non-specific binding of the enzyme labels on the membrane caused low
detection sensitivity[2]. Nowadays, sensitive DNA hybridization detection has become
increasingly important in the area of biomedical studies, clinical diagnosis and gene expression.
The polymerase chain reaction (PCR), which is one of the most important methods for DNA
detection, is developed with high sensitivity. However, the amplification of PCR includes a
thermal cycling process which is time consuming[3]. Thus, new DNA detection methods without a
thermal cycling process, for example, ion-sensitive field-effect transistor (ISFET) based DNA
detection[4] and gold electrodes with grafted self-assembled monolayer based detection[5] were
developed. Although these methods are rapid, the detection sensitivity was limited by the
electrical properties of metal or semiconductor electrodes. So it is of great necessity and
importance to develop a new sensing platform with high sensitivity to detect DNA hybridization
process.
Nanoporous alumina membranes, which have the honeycomb-like pore structures, are
formed by anodic etching of pure aluminum substrate in an oxalic, sulfuric or chromic aqueous

solution[6]. The pore geometry and morphology can be controlled by the electrolyte type and the anode voltage used. Nanoporous alumina membranes have relatively quick electrolyte transfer rate, increased surface affinity and reaction area compared with flat substrates, which can dramatically enhance the output sensing signals. They have been used for many biosensing applications including anti-cancer drug screening, cell behaviour monitoring, and food borne pathogen detection due to their well established fabrication process, low cost and high surface to volume ratio. The high surface reaction density allows large amounts of target molecules binding and dramatically increases the sensitivity.

The purpose of this study is to develop an optimal functionalized nanoporous alumina membrane based sensor for "label-free" DNA detection with gold nanoparticle amplification via impedance spectroscopy. The immobilization of ss-DNA on nanopore wall was achieved via silane-$(PEG)_{5000}$-NHS linker. The linker density was optimized to achieve the most efficient DNA immobilization efficiency. In the process of DNA detection, specific combination is very important to improve the detection accuracy. Poly (ethylene glycol) (PEG) chains can form a coating, which has the ability of reducing the amount of non-specific adsorption[7]. Furthermore, PEG derivative with functional groups, such as NHS esters and silane, can be used to immobilize DNA[8]. The silane-$(PEG)_{5000}$-NHS (Nanocs, New York, NY) can react with hydroxyl group of the sensor surface and the NHS group can covalently link with DNA. Thus, the silane-$(PEG)_{5000}$-NHS can be employed in the investigation of DNA detection based on nanoporous alumina to decrease non-specific adsorption. Moreover, DNA segments are so small that they need to be conjugated with nanoparticles for signal amplificaiton. Gold nanoparticles (AuNPs) have the advantages of biocompatibility and large surface areas for immobilization. They have been used to label proteins for cancer detection[9]. Also, they can bind to DNA with excellent electrochemical and optical properties[10], so they are applied to this study of DNA detection with nanoporous alumina membrane.

METHODS

Nanoporous membrane surface modification

Nanoporous alumina membranes with diameter of 100 nm were boiled in hydrogen oxide (H_2O_2) for 30 minutes to clean the membrane surfaces and generate -OH group. Then, they were washed in DI water with gentle shaking for 15 minutes and dried at 200 ℃ for 20 minutes. They were immersed in silane-$(PEG)_{5000}$-NHS (Nanocs Inc., USA) mixed with pure DMSO (Bio Basic Inc., Canada) for 2 hours at room temperature. DMSO could minimize hydrolysis of the NHS groups and preserve the NHS group reactive for ssDNA immobilization with the scheme shown in Figure 1.

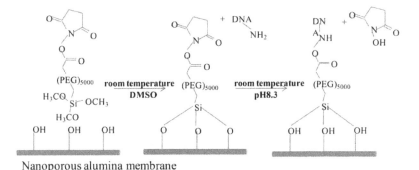

Nanoporous alumina membrane

Figure 1. Scheme of nanoporous alumina membrane modification

FITC labeled single stranded DNA was used for fluorescence detection and the DNA sequence was 5'-FITC-GTGTTTTCATCCAACAAACCTGCCC-3'. Amino modified DNA 5'-NH₂-CGACCAACCTAGCCTGCTATGATGT-3' was used as the probe to immobilize on the nanoporous alumina membrane surface for target DNA (5'-ACATCATAGCAGGCTAGGTTGGTCG-3') capture. DNA hybridization was done by adding the target DNA from PBS (pH 8.3) initially heating for an hour at 95^0C and slowly cooling down to room temperature[11].

Microdevice fabrication

A PDMS device with upper chamber and bottom chamber was designed and integrated with nanoporous alumina membrane for impedance detection (Figure 2). Sylgard 184 silicone elastomer base and curing agent with the ratio 10:1 were thoroughly mixed and poured into a mold. It was degassed for half an hour and put in oven for curing at $72\ ^0$C for 2 hours. When it became solid, the PDMS device was separated from the mold and integrated with nanoporous membrane.

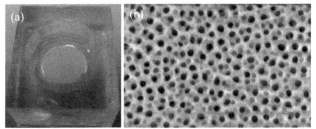

Figure 2. The detection device and structure of nanoporous alumina membrane

DNA-gold nanoparticle (AuNP) conjugation

AuNPs can be synthesized by reducing gold from chloroauric acid (HAuCl₄) by citrate[12]. Beakers and stir bar were washed thoroughly in aqua regia (mixture of HCl and HNO₃ with the ratio of 3:1) and rinsed in DI water. They were dried in oven prior to use. A mixture of HAuCl₄ (3 μL, 14.3 wt %) and DI water (10 ml) were boiled with vigorous stirring. Then, sodium citrate solution (1 ml, 1 wt %) was added to the boiling solution within one second. In a few minutes, the solution color changed from dark purple to deep red and it continued to boil for 15 minutes. The solution was left to cool down to room temperature by stirring and was filtered by a 0.8 μm membrane filter. The AuNP size and concentration were calculated from the UV absorbance data[13].

The detailed protocol of DNA-AuNP conjugation is based on previous study[14].Thiol modified DNA 5'-SH-ACATCATAGCAGGCTAGGTTGGTCG-3' was purchased from Life Technologies. It was treated with DTT diluted in PBS (pH 8.2) with the concentration 0.1 M for 30 minutes to cleave the disulfur linkage. The activated DNA was purified by gel columns (illustra MicroSpin G-25 Columns, GE Healthcare, UK) based on the manufacturer's instructions. The concentration of purified thiol modified DNA was estimated by UV absorbance. Immediately, it was added to the AuNP solution (100 μL, 3.8 nM DNA, 7.6 nM AuNP) standing for 24 hours at room temperature. Then, PBS (5 mM, pH 7.4) and sodium solution (0.1 M NaCl, 5mM NaH$_2$PO$_4$) were added to age for 16 hours. After that, it was centrifuged for 30 min at a high speed 13,200 rpm to remove excess DNA segments. The remaining red oily precipitate was washed by PBS solution. It was centrifuged for the second time and dispersed again in DI water for later use.

RESULTS AND DISCUSSION

To demonstrate the concentration effect of ssDNA immobilization on the nanoporous alumina membrane, qualitative and quantitative characterizations were performed using fluorescent microscopy. Fluorescence iamges of different DNA concentrations with 3% silane-(PEG)$_{5000}$-NHS are shown in Figure 3. Fluorescence images shown in Figure 3A, B, C, and D were with different DNA concentrations of 1μM, 3μM, 5μM and 7μM, separately. As the DNA concentration increased, fluorescence intensity increased with color changed from dark green to bright green.

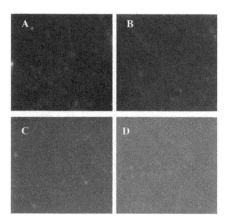

Figure 3. Fluorescence images of different ssDNA concentrations.

Fluorescence intensity analysis is shown in Figure 4. From the histogram, the average fluorescence intensity of the four different concentrations were 40.38, 43.47, 53.06 and 75.52, respectively. The fluorescence analysis results demonstrate that ssDNA are immobilized inside the nanopores and are available for the hybridization process in the next step. With the increasing silane concentration, more ssDNA were immobilized on the nanopores.

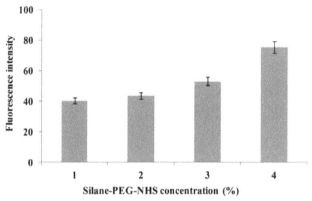

Figure 4. Fluorescence intensity of different DNA concentrations with 3% silane-$(PEG)_{5000}$-NHS

Nanoporous alumina membrane can act as a sensing platform for electrolyte impedance spectroscopy (EIS) analysis. The processes of ssDNA immobilization, and complimentary target DNA hybridization blocked ions passing through nanopores which could cause impedance change. The main parameters affecting signal amplitude including DNA concentration and, immobilization time and hybridization time have been analyzed. To increase detection sensitivity, AuNPs with diameter of 13 nm was chosen. The concentration of immobilized ssDNA was 3.8 nM. The complimentary target DNA concentration was 3.8 nM and AuNP (13 nm) concentration was 2.5 nM. Figure 5 shows spectrum before and after ssDNA immobilization, complimentary target DNA hybridization, and complimentary target DNA hybridization conjugated with AuNPs with the frequency range from 1 Hz to 100 Hz. During the experiments, membrane was washed three times gently by injecting PBS solution with a syringe at a rate of 5 µL/min.

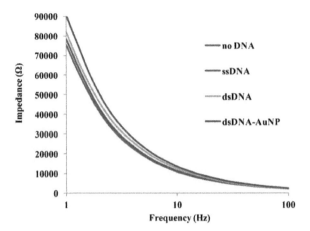

Figure 5. Impedance of nanoporous alumina membrane without DNA, single strand DNA, double strand DNA, and double strand DNA with gold nanoparticle.

At the low frequency around 1 Hz, the impedance amplitude showed the maximum change with the result (Figure 6). So, the impedance magnitude at 1 Hz was found to increase by 4.2%, 9.2% and 20.1% for ssDNA immobilization, dsDNA hybridization and AuNP amplification. The slight impedance amplitude increase with ssDNA immobilization demonstrated low non-specific adsorption of DNA on hydrolyzed PEG layer on membrane surface. The significant impedance amplitude increase around 20% after AuNP conjugation demonstrated efficient signal amplification.

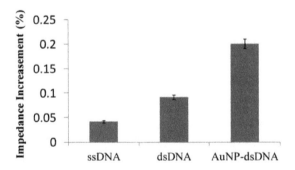

Figure 6. Relative impedance amplitude change for various cases compared with bare nanoporous membrane at 1 Hz.

CONCLUSION

An optimal functionalized nanoporous alumina membrane based sensor for "label-free" DNA detection has been developed. Using this nanoporous alumina membrane based device, DNA was successfully monitored via impedance spectroscopy. This portable PDMS device integrated with functionalized nanoporous alumina membrane with gold nanoparticle amplification has the potential for nucleic acid in field detection in the application of food screening and environmental monitoring.

FOOTNOTES
* Email: mo.yang@polyu.edu.hk

REFERENCES
[1]A. Behzad-Behbahani, A. Mafi-Nejad, S. Tabei, K. Lankarani, A. Torab and A. Moaddeb, Anti-HBc & HBV-DNA detection in blood donors negative for hepatitis B virus surface antigen in reducing risk of transfusion associated HBV infection, *Indian J. Med. Res.*, **123**, 37-42 (2006).
[2]K. A. DeFillipo and M. L. Grayeski, Flow-injection chemiluminescent method for an enzyme-labelled DNA probe, *Anal. Chim. Acta.*, **249**, 155-62 (1991).
[3]S. Kambhampati and P. Smith, PCR primers for the amplification of four insect mitochondrial gene fragments, *Insect Mol. Biol.*, **4**, 233-36 (1995).
[4]P. Bergveld, Thirty years of ISFETOLOGY: What happened in the past 30 years and what may happen in the next 30 years, *Sensors Actuators B: Chem.*, **88**, 1-20 (2003).
[5]S. D. Vernon, D. H. Farkas, E. R. Unger, V. Chan, D. L. Miller, Y. Chen, G. F. Blackburn and W. C. Reeves, Bioelectronic DNA detection of human papillomaviruses using eSensor™: a model system for detection of multiple pathogens, *BMC Infectious Diseases*, **3**, 12 (2003).
[6]W. Lee, R. Ji, U. Gösele and K. Nielsch, Fast fabrication of long-range ordered porous alumina membranes by hard anodization, *Nature Materials*, **5**, 741-47 (2006).
[7]R. Michel, S. Pasche, M. Textor and D. G. Castner, Influence of PEG architecture on protein adsorption and conformation, *Langmuir*, **21**, 12327-32 (2005).
[8]D. J. Schlingman, A. H. Mack, S. G. Mochrie and L. Regan, A new method for the covalent attachment of DNA to a surface for single-molecule studies, *Colloids and Surfaces B: Biointerfaces*, **83**, 91-95 (2011).
[9]A. de la Escosura☐Muñiz and A. Merkoçi, A Nanochannel/Nanoparticle☐Based Filtering and Sensing Platform for Direct Detection of a Cancer Biomarker in Blood, *Small*, **7**, 675-82 (2011).
[10]A. Merkoçi, M. Aldavert, G. Tarrasón, R. Eritja and S. Alegret, Toward an ICPMS-linked DNA assay based on gold nanoparticles immunoconnected through peptide sequences, *Anal. Chem.*, **77**, 6500-03 (2005).
[11]I. Vlassiouk, A. Krasnoslobodtsev, S. Smirnov and M. Germann, "Direct" Detection and Separation of DNA Using Nanoporous Alumina Filters, *Langmuir*, **20**, 9913-15 (2004).
[12]K. C. Grabar, R. G. Freeman, M. B. Hommer and M. J. Natan, Preparation and characterization of Au colloid monolayers, *Anal. Chem.*, **67**, 735-43 (1995).
[13]W. Haiss, N. T. Thanh, J. Aveyard and D. G. Fernig, Determination of size and concentration of gold nanoparticles from UV-vis spectra, *Anal. Chem.*, **79**, 4215-21 (2007).
[14]J. K. Wong, S. P. Yip and T. M. Lee, Silica☐Modified Oligonucleotide– Gold Nanoparticle Conjugate Enables Closed☐Tube Colorimetric Polymerase Chain Reaction, *Small*, **8**, 214-19 (2012).

Author Index

Author Index

Seto, J., 75
Shokrgozar, M. A., 113

Tas, A. C., 91
Teoh, S. H., 179
Thian, E. S., 179
Tseng, H., 135

Wang, Z., 135

Yang, M., 191
Yang, W., 105
Ye, W. 191
Yekta, B. E., 113
Yoon, H.-J., 149
Youssefi, A., 113

Zhang, Z. Y., 179